OXYGEN TRANSPORT TO TISSUE XXVIII

ADVANCES IN EXPERIMENTAL MEDICINE AND BIOLOGY

Recent Volumes in this Series

Volume 591
SOMATIC CELL NUCLEAR TRANSFER
Edited by Peter Sutovsky

Volume 592
REGULATORY MECHANISMS OF STRIATED MUSCLE CONTRACTION
Edited by Setsuro Ebashi and Iwao Ohtsuki

Volume 593
MICROARRAY TECHNOLOGY AND CANCER GENE PROFILING
Edited by Simone Mocellin

Volume 594
MOLECULAR ASPECTS OF THE STRESS RESPONSE
Edited by Peter Csermely and Laszlo Vigh

Volume 595
THE MOLECULAR TARGETS AND THERAPEUTIC USES OF CURCUMIN IN HEALTH AND DISEASE
Edited by Bharat B. Aggarwal, Yung-Joon Surh and Shishir Shishodia

Volume 596
MECHANISMS OF LYMPHOCYTE ACTIVATION AND IMMUNE REGULATION XI
Edited by Sudhir Gupta, Frederick Alt, Max Cooper, Fritz Melchers and Klaus Rajewsky

Volume 597
TNF RECEPTOR ASSOCIATED FACTORS (TRAFs)
Edited by Hao Wu

Volume 598
CURRENT TOPICS IN INNATE IMMUNITY
Edited by John D. Lambris

Volume 599
OXYGEN TRANSPORT TO TISSUE XXVIII
Edited by David Maguire, Duane F. Bruley and David K. Harrison

David J. Maguire
Duane F. Bruley
David K. Harrison
(Eds.)

Oxygen Transport to Tissue XXVIII

Springer

David Maguire
Genomics Research Centre
Griffiths University
Parklands Drive
Southport 4217, QLD
Australia
d.maguire@griffith.edu.au

Duane F. Bruley
University of Maryland
1000 Hilltop Circle
Baltimore County (UMBC)
Baltimore, MD 21250
USA
bruley@umbc2.umbc.edu

David K. Harrison
University Hospital of North Durham
North Road
Durham DH1 5TW
UK
d.k.harrison@ncl.ac.uk

Proceedings from the 33rd Annual Meeting of the International Society on Oxygen Transport to Tissue (ISOTT) held in Brisbane, Australia, August 28-September 2, 2005.

Library of Congress Control Number: 2007920486

ISBN-13: 978-0-387-71763-0
e-ISBN-13: 978-0-387-71764-7

Printed on acid-free paper.

10 9 8 7 6 5 4 3 2 1

springer.com

INTERNATIONAL SOCIETY ON OXYGEN TRANSPORT TO ISSUE 2005-2006

The International Society on Oxygen Transport to Tissue is an interdisciplinary society comprised of approximately 260 Members worldwide. Its purpose is to further the understanding of all aspects of the process involved in the transport of oxygen from the air to its ultimate consumption in the cells of various organs of the body.

The annual meetings bring together scientists, engineers, clinicians and mathematicians in a unique international forum for the exchange of information and knowledge, the updating of participants on the latest developments and techniques, and the discussion of the controversial issues within the field of oxygen to tissue.

Examples of areas in which members have made highly significant contributions include electrode techniques, spectrophotometric methods, mathematical modeling of oxygen transport, and the understanding of local regulation of oxygen supply to tissue and fluorocarbons/ blood substitutes.

Founded in 1973, the society has been the leading platform for the presentation of many of the technological and conceptual developments within the field, both at meetings themselves and in the proceedings of the society. These are currently published by Springer in the *Advances in Experimental and Biology* series, which is listed with an impact factor in the Science Citation Index.

Officers

David J. Maguire (Australia)	President
Giuseppe Cicco (Italy)	Past President
Kyung A. Kang (USA)	President Elect
Oliver Thews (Germany)	Secretary
Peter E. Keipert (USA)	Treasurer
Duane F. Bruley (USA)	Chairperson, Knisely Award Committee

Executive Committee

Chris Cooper (UK)
Jerry D. Glickson (USA)
Per Liss (Sweden)
Valentina Quaresima (Italy)
Peter Vaupel (Germany)

Local Committee

Dr Russell Addison
Nicholas Lintell
Harald Oey
Jan Shah
Cathy Hsieh

New Member of Executive Committee Elected ISOTT 2005

Fahmeed Hyder (USA)
Paul Okunieff (USA)
Akitisho Seiyama (Japan)
Christpher B. Wolff (UK)

Awardees 2005

Melvin H. Knisely Award: Nicholas Lintell (Australia)
Dietrich W. Lübbers Award: Charlotte Ives (UK)
Britton Chance Award: James Lee (USA)
Duane F. Bruley Awards: Robert Bradley (UK), Harald Oey (Australia), Kathy Hsieh (Australia), Jan Shah (Australia)

PANEL OF REVIEWERS

Duane Bruley, University of Maryland Baltimore County, USA
David Delpy, University College London, UK
Clare Elwell, University College London, UK
Jerry Glickson, University of Pennsylvania, Philadelphia, USA
David Harrison, University Hospital of North Durham, UK
Kyung Kang, University of Louisville, USA
Joe LaManna, Case Western Reserve University, USA
Paul Okunieff, University of Rochester, USA
Harold Swartz, Dartmouth Medical School, Hanover, USA
Oliver Thews, University of Mainz, Germany
Maureen Thorniley, University of Manchester, UK
Christopher Wolff, St Thomas's Hospital, London, UK

PREFACE

At the beginning of a Southern Hemisphere Autumn, the ISOTT tribe assembled for its annual corroboree on the banks of the Brisbane River in Australia for five days of exciting science. Some of the tribe had travelled from as far as Sweden and other parts of Europe; many had spirited themselves across the Pacific Ocean from the Americas. Yet another small contingent had paddled down from Japan. All came with message sticks from their camp-sites.

The conference opened with an outdoor barbecue at the Southbank precinct, with the stars of the Southern Cross painting the river on one side and the artificial beach on the other; a perfect venue to meet old friends and to welcome new ones as a cool breeze wafted over our fires.

For the next five days there was the usual focus on oxygen; measurement techniques, its role in cancer and other diseases that beset us mortals and the biochemistry and physiology of this small vital molecule. It still astounds many that we have not yet learnt everything there is to know about this valuable friend, albeit a dangerous foe.

Time was found for us to visit Australia Zoo and stand in awe before exhibits of Australia's wild-life; great vicious creatures unchanged from the age of the dinosaurs, poisonous snakes and spiders, and gentle marsupials that posed for our digital cameras.

All too soon, the final lecture was delivered, the banquet was upon us and the prizes awarded. Time arrived then to say our farewells, each delegate carrying home a boomerang to remind us to attend the next meeting of ISOTT, and many more after that. The editors thank all participants in this conference as well as those who worked behind the scenes to ensure that things ran smoothly. We especially want to acknowledge the assistance of our technical editors, Larraine Visser and Eileen Harrison in the preparation of this volume.

TABLE OF CONTENTS

ANTICOAGULANT BLOOD FACTOR DEFICIENCIES (PROTEIN C)

Duane F. Bruley*

Abstract: Anti-coagulant proteins are essential to maintain blood hemostatis for the supply of oxygen and nutrients to tissue cells and for the removal of toxic by-products from metabolism. Hereditary or acquired deficiencies of Protein C, Protein S, or Anti-thrombin III can lead to disease states such as deep vein thrombosis (DVT) with the possibility of producing lung emboli. Phenomena named Factor V Lieden can produce a similar pathologic condition.

Anti-coagulant deficiencies, including Factor V Lieden, are HIDDEN blood conditions that can allow blood clot development, especially with trauma to the tissue and circulatory system. It is proposed that all children between ages twelve to fourteen be checked hereditary deficiencies and Factor V Lieden complications. This would require the development of inexpensive assay equipment12.

The present research focuses on the low cost production of Zymogen Protein C via purification from blood plasma Cohn Fraction IV-1. This process is difficult due to the several Homologous Vitamin K dependent proteins in the blood coagulation cascade. Traditional chromatography (ion exchange) cannot achieve the desired separation. Some more exotic technologies are very expensive so our work proposes to use Immobilized Metal Affinity Chromatography (IMAC). It is hoped to produce a lower cost product that can be used prophylactic ally to treat Protein C deficiencies and possibly other coagulation problems.

1. INTRODUCTION

Pathologic biochemical and biomechanical events in living tissue and organs can result in oxygen reduction and thus starvation of cells leading to tissue death. With this in mind, blood hemostasis is a major consideration in health and disease.[1] Ever since man has become aware of oxygen transport in the microcirculation we have been looking for procedures, technologies and interventions to maintain blood hemostasis and thus normal

* Duane F. Bruley, Ph.D., P.E., Synthesizer Inc, Ellicott City, MD, College of Engineering, University of Maryland Baltimore County, Baltimore, MD 21250

oxygen supply to cells. A complicating factor in essentially all disease states is the abnormal clotting of blood where the particulate matter sticks to each other and to the endothelial cells. This phenomena decreases or stops blood flow and therefore oxygen and nutrient supply to tissues and prevents the removal of toxic metabolic byproducts.

A very important medical consideration is the availability of essential blood factors integral to the clotting cascade and/or biochemicals that can be used to control the clotting process for healthy functioning of tissue. This paper focuses on abnormal blood clotting, the pharmacological intervention to help prevent abnormal blood clotting, dissolution of clots that have formed, as well as strategies to eradicate the known pathological conditions leading to clot formation. It is well established that inflammation plays a role *via* cell membrane function that is a consideration in the sticking together of particulate matter such as red cells, white cells, platelets, endothelial cells, *etc*. It is well known that many disease states such as AIDS, sepsis, myocardioinfarction, diabetes, and cancer can result in catastrophic events that lead to death *via* the plugging of blood vessels and capillaries.

Great strides have been made in understanding the pathogenetic role of erythrocyte and thrombocyte aggregation in thrombotic disease.[2,3] Two somewhat different mechanisms lead to a similar result of tissue death. The two major phenomena include erythrocyte aggregation and platelet aggregation. These mechanisms have required intense research to determine optimal anticoagulant and antithrombotic biological and pharmacological agents to counteract these processes. A huge industry has evolved in the development of drugs to maintain blood hemostasis and regulate the blood clotting cascade to prevent hemophilia and thrombosis. This is an extremely challenging objective, since there are so many different degrees and types of aggregation. Studies of the underlying mechanisms can help identify the abnormal biomechanical and biochemical forces holding together the aggregated cells, thus leading to an understanding that will help us to design strategies that will prevent tissue damage and death.

Included in this search for the optimal anticoagulant, antithrombotic, and anti-inflammatory drug are the natural blood factors in human blood that link together through a complex network to maintain normal blood flow and tissue oxygenation. Our work focuses on protein C[4], the pivotal anticoagulant with antithrombotic, anti-inflammatory and anti-apoptotic[5,6] characteristics. At this time I am proposing that the protein C zymogen could represent the optimal anticoagulant for prophylactic use in the case of anticoagulant deficiencies because its function is local and it does not lead to bleeds and other dangerous side effects that all other known anticoagulants possess.

2. EARLY STUDIES AND THE IMPORTANCE OF NATURAL ANTI-COAGULANTS

Dr. Melvin H. Knisely has been cited as the first scientist to observe *in vivo* the sticking together of particulate matter of blood, namely erythrocytes, leucocytes, and platelets. His original development of the quartz rod crystal technique for illuminating and magnifying the microcirculation in the eye allowed him to observe "blood sludging" in over 100 disease states (Figure 1). This was a very significant discovery, because man has searched since the beginning of time for the answer to what happens inside of the body as death approaches. Figure 2 is one of his photographs taken through the thin

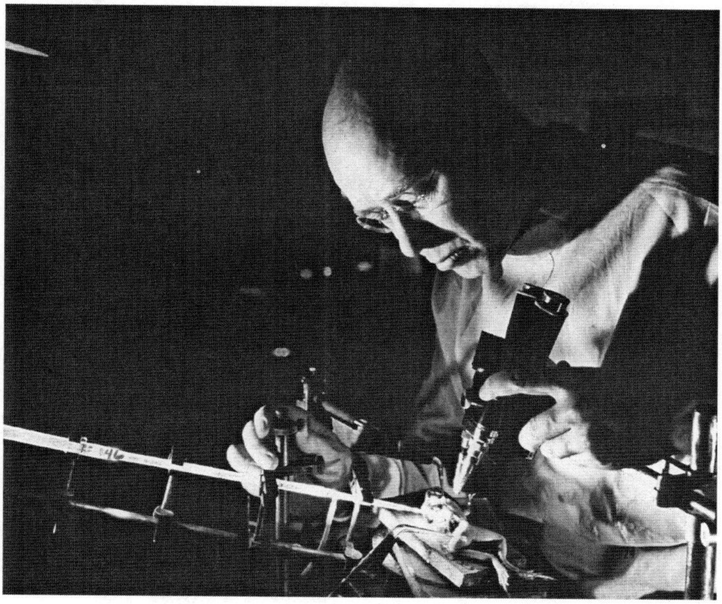

Figure 1. Dr. Melvin H. Knisely observing *in vivo* sludging: Using a quartz rod to conduct light to a frog's organs, Dr. Knisely bends toward the microscope to observe the circulating blood. Fritz Goro/ Getty Images.

tissue of the belly of a frog, illustrating the microcirculation at death resulting from traumatic shock.

Human health is highly dependent upon blood protein levels to maintain blood hemostasis. To prevent thrombotic events in the circulatory system there are at least three very important natural proteins. These high molecular weight therapeutic proteins include protein C, protein S, and antithrombin III. The pivotal anticoagulant in human blood is considered by many to be protein C. However, deficiencies in any of these three proteins can lead to thrombosis—the most prevalent being deep vein thrombosis (DVT)[7]—which can result in clot emboli that can plug the microcirculation in the lungs, leading to death. Protein S is a cofactor of protein C and is necessary to activate protein C at a thrombin/thrombomodulin complex on the endothelial cells thus allowing activated protein C to deactivate factor Va and factor VIIIa. Antithrombin III works with a different mechanism to prevent clot formation. Anticoagulant deficiency can be a hereditary complication or can be acquired *via* disease states, trauma, and other intrusions. For example, it is common for protein C levels to drop during pregnancy, after birth, and in advance sepsis. These situations can lead to blood clotting and Deep Vein Thrombosis (DVT) similar to hereditary deficiencies in the anticoagulants. In the case of pregnancy it might be prudent to administer a zymogen protein C product for at least the ninth month of pregnancy and one month post partum.

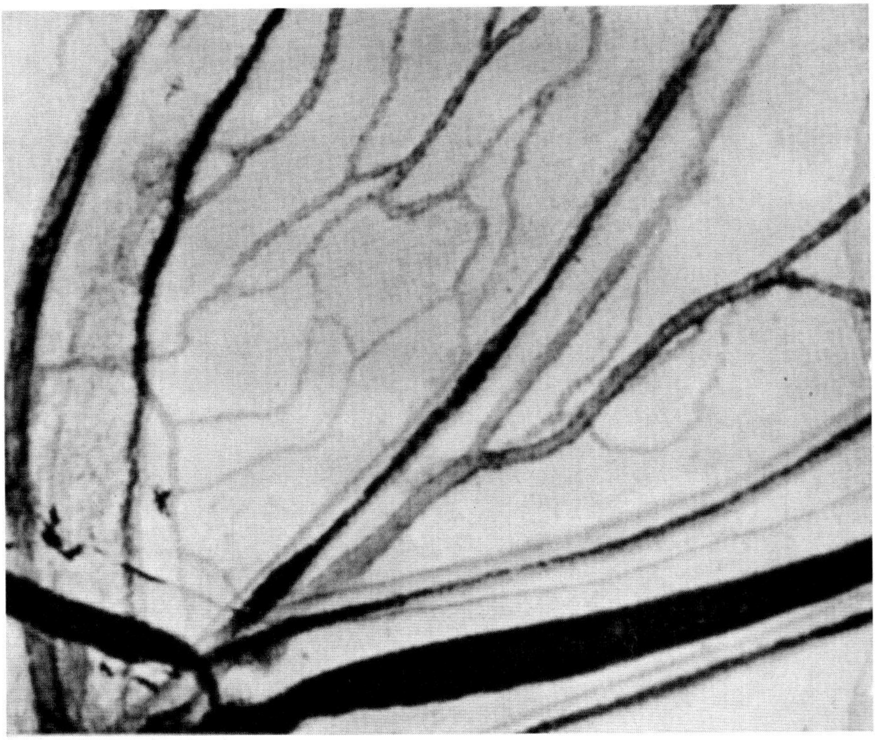

Figure 2. Microcirculation in frog's belly at death[7].

Another hereditary and pathological state exists in many people labeled the Factor V Leiden complication. This condition is a variant of the protein Factor V which can lead to blood that has an increased tendency to clot. It is the most common blood clotting disorder; as high as 5% of the population carry genes for Factor V Leiden. The most probable end result would once again be the formation of deep vein thrombosis (DVT) with the high risk of emboli reaching the lungs.

Since anticoagulant deficiencies and Factor V Leiden are hidden complications, the writer recommends that *all children between the ages of twelve to fourteen should have a blood test that determines hereditary deficiencies and abnormalities that could lead to DVTs and strokes*, particularly those children with family histories of blood clotting complications. The results of these tests would allow parents to be alert for potential problems and to provide a lifestyle for the child that would reduce the risks of blood clots[8].

3. THE PIVOTAL ANTICOAGULANT – PROTEIN C

Our research targets developing reduced-cost processing for the production of blood factors. As stated previously, protein C deficiency can be either hereditary or acquired with present treatments utilizing heparin and/or coumadin to prevent clotting.

Homozygous deficiencies are present in 1 in 400,000 births and result in almost immediate death of the child, while the heterozygote population is approximately in 1 in 300 people. However, a typical number would be 1 in 16,000 for those who are symptomatic and experience major complications. To eliminate the potential catastrophic side effects of both heparin and coumadin, research is underway to produce a low-cost anticoagulant with essentially no side effects. Protein C has anticoagulant benefits; in addition it is anti-inflammatory, antithrombotic and anti-apoptotic and it circulates in the blood as a zymogen that functions on site and on demand with no bleeding side effects.

There are three known sources of protein C which include recombinant DNA technology/bioreactor production, transgenic animal milk, and blood plasma. This research focuses on the recovery and purification of protein C from blood plasma Cohn Fraction IV-I[9]. The biggest engineering challenge for protein C production from the three sources is reducing process cost for general availability of the product. For instance the product Xigris™ (activated protein C), produced by Eli Lilly, and is available for treatment of advanced sepsis. However, the cost is thousands of dollars per dose and therefore is only used as a last resort. Ceprotin™, produced by Baxter Labs, is a blood plasma cocktail that has been successful in treating protein C deficiencies. However, it is produced using immunoaffinity chromatography at great expense.

In an attempt to produce zymogen protein C in large quantities at low cost we are exploring the use of immobilized metal affinity chromatography for the separation and purification of a protein C cocktail from blood plasma fraction IV-1. This product could be used prophylactically to treat protein C deficiency to prevent abnormal blood clotting. This is an extremely difficult separation because of the homologous nature of high molecular weight blood factors in the clotting cascade that are found in the source material. It is particularly important to reduce the levels of prothrombin (factor II) because of its long half-life *in vivo* and its relatively high concentration in blood. It has been demonstrated that metal affinity chromatography has a high specificity and therefore is capable of separating protein C from prothrombin.[10] A process incorporating immobilized metal affinity chromatography (Figure 3) is presently under development with strong indications that it will be capable of producing a relatively inexpensive zymogen protein C product.[11,12] Further studies are being completed to optimize the operating conditions to achieve the least expensive bioprocess possible. The thermodynamic parameter, Gibbs Free Energy, is being studied in an attempt to determine the most effective concentration of imidazole for Protein C elution and therefore obtain a clean cut for Protein C *versus* prothrombin and other undesirable blood factors.

Another approach using mini-antibodies produced from recombinant *E. coli* is being studied.[13] This technology has promise for producing low-cost protein C. However, a cost comparison with metal affinity chromatography will not be possible until both processes are completely finished and the resulting products can be compared on the same basis.

4. ACKNOWLEDGEMENTS

The author wishes to thank Ms. E. Eileen Thiessen for her help in developing Power Point slides, typing this manuscript and for providing important feedback on the material.

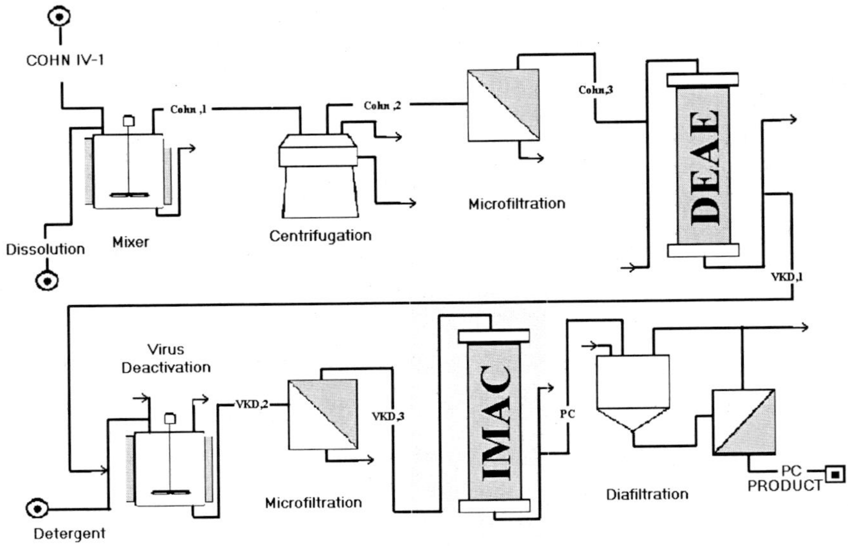

Figure 3. Tentative Downstream Bioprocess for Protein C Production[10].

5. REFERENCES

1. Bicher, H. I., Blood Cell Aggregation in Thrombotic Processes, *A Mongraph in The Bannerstone Division of American Lectures in Hematology,* Ed. Walter H. Seegers, **835** (1972).
2. Frojmovic, M., G. Nash, S.L. Diamond, "Cell Aggregation and Cell Adhesion in Flow," *Thromb. Haem.* **87**:771 (2002).
3. Diamond, S. L., P. Tandon, D. Schmidtke, I. Laurenzi. "Cellular Aggregation in Blood Flow." *Comments Theoretical Biol.* **8**:1 (2000).
4. Bruley, D. F., and W. N. Drohan, "Protein C and Related Anticoagulant," Advances in Applied Biotechnology Series, **11** (1990).
5. Cheng, T., Liu, D., Griffin, J. H. *et. al.*, "Activated protein C blocks p53-mediated apoptosis in ischemic human brain endothelium and is neuroprotective," *Nature Medicine*, **9**:338 (2003).
6. Mosnier, L. O. and J. H. Griffin, "Inhibition of staurosporine-induced apoptosis of endothelial cells by activated protein C requires protease-activated receptor-1 and endothelial cell protein C receptor," *Biochem. J.*, **373**:65 (2003).
7. http://www.preventdvt.org/coalition.info/
8. Tang, L., Ren, Y.J., Hong, B. and Kang, K.A., "A Fluorophore-mediated, Fiber-optic, Multi-analyte, Immuno-sensing System for the Rapid Diagnosis and Prognosis of Cardiovascular Diseases." Journal of Biomedical Optics, 11(2), 021011, March/April 2006.
9. Chen, T. American Red Cross, Jerome Holland Laboratories, Rockville, Maryland, USA.
10. Wu, H., and D. F. Bruley, "Chelator, metal ion and buffer studies for protein C separation," *Comparative Biochemistry and Physiology Part A,* **132**:213 (2002).
11. Wu, H., and D. F. Bruley, "Process Studies for Protein C Separation," *Oxygen Transport to Tissue XXVI* in Advances in Experimental Medicine and Biology, *in press*.
12. Thiessen, E. E., and D. F. Bruley, "Theoretical Studies of IMAC Interfacial Phenomena for the Production of Protein C," *Oxygen Transport to Tissue XXV* in Advances in Experimental Medicine and Biology, **540**:183 (2003).
13. Korah, L.K. and K. A. Kang, "Preliminary Study for the Protein C Purification Using Mini-Antibodies Produced from Recombinant E-coli," M. S. Thorniley, D. K. Harrison, and P. E. James, *Oxygen Transport to Tissue XXV* in Advances in Experimental Medicine and Biology, Kluwer Academic/Plenum Publishers, **540**:171 (2003).

HEMORHEOLOGICAL ASPECTS IN THE MICROVASCULATURE OF SEVERAL PATHOLOGIES

Giuseppe Cicco and Sebastiano Cicco*

Abstract: We evaluated morphological changes in several pathologies using computerized videocapillaroscopy, and related hemorheological patterns using the laser assisted optical rotational red cell analyzer (LORCA). In addition, tissue oxygenation was measured using two oximeters with Combi sensors (Periflux 5000, Perimed).

The study included four groups of patients (pts) that were compared with a control group. **Group A** *Controls* (n=25: 15 males [M] and 10 females [F] aged 36±3 years); **Group B** *Diabetic* pts n=32 (IDDM pts n=20: 12 M and 8 F aged 43±4 years; NIDDM pts n=12: 6 M and 6 F aged 45±3 years); **Group C** *Glaucoma* pts n=30 (16 M and 14 F aged 42±5 years); **Group D** *Liver failure* pts n=6 (3 M and 3 F aged 44±5 years); **Group E** *Hypertensive* pts n=50 (smokers n=28: 12 M and 16 F aged 40±4 years, and non-smokers n=22: 12 M and 10 F aged 38±3 years). In all patients hemorheological measurements were made using the LORCA (including red blood cell [RBC] deformability and aggregability), morphology was evaluated using computerized videocapillaroscopy (magnification 200 x), and transcutaneous oxygen partial pressure measurements ($TcpO_2$) were made with the Periflux 5000.

In patients with diabetic microangiopathy: the capillary loops in 50% (16/32) of these pts showed formations such as 'deer horns', 72% (23/32) showed formations such as 'elephant nose', and in 45% (14/32) formations such as a 'cork screw'; in diabetics with POAD an important capillary rarefaction was found in 26% (9/32) of the pts. In glaucoma patients, in 84% (25/30) we observed 'capillary meandering' and images such as 'a comb'. In patients with more complicated pathology capillary rarefaction was found in 70% (21/30) of the patients. An improvement in the perfusion of non-functional loops was found in deceased patients who had suffered liver failure one week after liver transplantation in 90% (5/6) of the studied cadavers. In non-smoking hypertensives morphological changes were found in 25% (6/22) of the patients, and in hypertensive smokers in 47% (13/28). RBC deformability was detected using LORCA and expressed

* **C.E.M.O.T.** – Intedipartimental Center of Research on Hemorheology, Microcirculation, Oxygen Transport to tissue and non invasive optical technologies University of Bari, Policlinico, P.za G. Cesare, 11 70124 Bari, Italy.Telefax 0039 080 5478156 email gcicco.emo@tiscali.it .

as the Elongation Index (EI), and RBC aggregability was detected using LORCA and expressed in t½ (seconds) indicating the RBC aggregability peak. Group A controls: EI 0.59±0.02; t½ 3±1 sec; Group B: IDDM EI 0.55±0.01; t½ : 2±0.5 sec $p<0.05$; NIDDM EI 0.56±0.01; t½ 2±0.2 sec $p<0.04$; Group C glaucoma: EI 0.56±0.01; t½ 2±0.3 sec $p<0.05$; Group D liver failure: EI 0.56±0.02; t½ 2±0.4 sec $p<0.03$; Group E hypertensives: smokers EI 0.56±0.02; t½ 2±0.6 sec $p<0.04$; non-smokers EI 0.57±0.02; t½ 2±0.6 sec $p<0.04$ compared with controls. We also measured the $TcpO_2$ at the dorsum of the right foot as a standard site representing peripheral control of microvasculature perfusion. Group A 96±11 mmHg; Group B IDDM 74±9 mmHg $p<0.05$; NIDDM 76±8 mmHg $p<0.05$; Group C glaucoma 75±9 mmHg $p<0.05$; Group D liver failure 69±6 mmHg $p<0.05$; Group E hypertensives: smokers 70±5 mmHg $p<0.05$, non-smokers 77±9 mmHg $p<0.05$ compared with controls.

This study presents an interesting and complete methodology to evaluate the microcirculation in different pathologies that induce changes in the microvasculature.

1. INTRODUCTION

The first to describe the circulatory system in detail was Sir William Harvey (1628; Fig.1); however, he did not explain the connection between the venous and the arterial system in his *De Motu Cordis*. In Italy the studies of Marcello Malpighi (1661; Fig. 2) have been very important, particularly his *De pulmonibus Observationes Anatomicae*. In his studies Malpighi described preliminary but nevertheless very important observations about the capillary system.

In Holland, Antoni van Leeuwenhoek (1675) devised a prototype instrument to directly study red blood cells (RBC): i.e. a microscope. In a letter to the Royal Scientific Society of London (1686) he reported his observations on RBC in the narrow capillaries described by Malpighi [1]. Subsequently, many other researchers made various scientific contributions. For example, in 1895 the 'Starling Law' described the oncotic and osmotic pressure in the filtration mechanism of vessel-tissue absorption; Starling observed that the contraction energy of the myocardiocytes is directly related to their initial length [2].

T. Lewis (1917), A. Krogh (1922) and Zweifach (1940) marked the birth of scientific knowledge on the microcirculation. In subsequent years studies were published on hemorheology and blood viscosity (L. Dintenfass 1971, T. Di Perri 1979, S. Forconi 1987, H. Meiselman 1989, M.R. Hardeman 1991); as well as on microhemodynamics (R. Del Guercio 1986), and on ischemia/reperfusion, especially during organ transplantation (G. Cicco 2003).

Today molecular and cellular biology are able to explain the relationships between the endothelium and the blood stream, leukocyte rolling and adhesion.

Endothelial functions are not included in the normal balance of endothelial cell stimulation and/or inhibition and are able to favour circulatory diseases. It then became interesting to study the hemorheological and morphological changes in tissue oxygenation in the microcirculation of various patients: including diabetics, hypertensives, patients suffering from lipoidoproteinosis, those with peripheral arterial occlusive disease (PAOD), with systemic pathologies such as systemic lupus erythematosus (LES), progressive systemic sclerosis (PSS= scleroderma), those undergoing organ transplantation (liver, kidney), and those suffering from primary open angle glaucoma (POAG).

Fig. 1. Sir William Harvey.

Fig. 2. Marcello Malpighi.

2. AIM

The aim of this study was to evaluate the morphological changes in several pathologies using computerized videocapillaroscopy (Mitsubishi, Videocap, Japan: magnification 200 x), the relative hemorheological patterns using the laser assisted optical rotational red cell analyzer (LORCA), and tissue oxygenation using two oximeters with combi sensors to measure PO_2 and PCO_2 (Periflux 5000, Perimed).

3. MICROCIRCULATION

According to Chambers and Zweifach (1940) the microcirculation is organized into many units: artery, capillary, arteriolar-venular anastomosis (not always present), veins and perivascular areas. The arteriolar-capillary joints and preference channels are very important. A simple capillary classification was made according to Bunnet (1958) and Rodhin (1973).

A) Related to the basilemma characteristics:
   1) type A continuous basilemma
   2) type B discontinuous basilemma (sinusoids)

B) Related to the endothelium characteristics:
   1) type 1 continuous and thick endothelium (subcutaneous tissues)
   2) type 2 continuous and thin endothelium (skeletal and smooth muscles, lungs, fat, central and peripheral nervous system)
   3) type 3 fenestrated endothelium (endocrine and exocrine glands)
   4) type 4 discontinuous endothelium (liver spleen, bone marrow)

C) Related to the perivascular tissues:
   1) type α evident perivascular tissues
   2) type β detached perivascular tissues

4. MATERIAL AND METHODS

In the present study four groups of patients were compared with a control group: Group A Controls, Group B Diabetes, Group C Glaucoma, Group D Liver Failure, and Group E Hypertension. All patients underwent a) hemorheological measurements using the LORCA which measures RBC deformability and RBC aggregability; b) morphological evaluation using computerized videocapillaroscopy (Videocap, Mitsubishi, Japan) i.e. a microscope/video imaging technique with a magnification of 200 x; and c) transcutaneous oxygen partial pressure measurement ($TcpO_2$) (Periflux 5000, Perimed). In addition, in Group C (30 glaucomatous patients) we also measured the RBC surface acetylcholinesterase (ACE) and the cytosolic calcium as hemorheological patterns during treatment of this pathology.

4.1 Videocapillaroscopy

Many pathologies are known to have related capillary abnormalities, e.g. endocrine, psychiatric, cardiovascular, and rheumatic diseases (Table 1).

Table 1. Pathologies often with related capillary abnormalities.

1. **Endocrine diseases**
   - Diabetes
   - Hypothyroidism
   - Addison's disease

2. **Psychiatry**
   - Schizophrenia
   - Epilepsy
   - Oligophrenia

3. **Cardiovascular diseases**
   - Arterial hypertension
   - Right heart failure

4. **Rheumatic diseases**
   - Progressive systemic sclerosis (PSS)
   - Dermatomyositis
   - Eosinophile fascitis
   - Mixed connective phlogosis
   - Raynaud's syndrome and disease
   - Rheumatic fever
   - Systemic lupus erythematosus (LES)
   - Psoriatic arthritis
   - Cryoglobulinemia

Videocapillaroscopy enables the study of vessels in the microcirculation providing information on their length, number, tortuosity, ectasia, stenosis, etc. [3]. Recently our Center has been using computerized videocapillaroscopy (Videocap, Mitsubishi, Japan; magnification 200 x) to evaluate the microvasculature throughout the body.

We generally follow the Lee Classification [3]:

Level 0	no avascular area
Level 1	1 or 2 avascular zones with small dimensions
Level 2	more than 2 avascular areas with extended dimensions
Level 3	confluent avascular zones (desertification).

In capillaroscopy the morphologic or static parameters (Table 2) and the functional or dynamic (Table 3) parameters are very interesting.

Table 2. Morphological or static parameters.

- Visibility
- Loop morphology
- Loop orientation
- Capillary density
- Loop length
- Loop diameter
- Loop-dermal papilla distance
- Loop tortuosity
- Ectasia and microaneurism
- Megacapillaries
- Avascular zones
- Neoangiogenesis
- Microbleedings

Table 3. Functional or dynamic parameters.

Capillary flow:

- Continuous or ribbon-like
- Granular or with microaggregates
- Intermittent or with plasma gaps
- With or without stops

Capillaroscopy can be used in clinical practice (Table 4) as well as in various fields of research (Table 5) [4].

Table 4. Capillaroscopy in clinical practice.

- Angiology (POAD)
- Phlebology
- Rheumatology (PSS, LES)
- Cardiology (hypertension)
- Vascular surgery
- Dermatology
- Surgery
- Oncology
- Endocrinology (diabetes, hypothyroidism)
- Psychiatry (schizophrenia, epilepsy, oligophrenia)

Table 5. Capillaroscopy in scientific research.

- Mapping of the regional architecture of the microvessels.
- Response of the regional microvessels to physico-mechanical, hemodynamic, neuroendocrine, postural and other stimulations.
- Response of the microvessels to pharmacological stimuli as well as to endogenous and exogenous stimulations.
- Monitoring of the efficacy of drugs.
- Study of microcirculation-tissue relationships in patients of different ages
- Regional dynamic hemorheology.

Many capillary morphological changes have been found during routine use of videocapillaroscopy:

1) normal loops, such as 'hairpins'
2) tortuous loops, such as 'corkscrews, candelabra, glomerulus'
3) branched loops, such as 'deer horns, shamrock' (Fig. 3)
4) arborescent loops, such as 'thicket, claw'
5) loops such as 'cactus'
6) loops such as 'shoal of fishes' or 'elephant nose' (Fig. 4) in diabetes
7) apical or lateral small ectasia
8) capillary loop enlargement (in the efferent loop or in the apical zone)
9) megacapillary in 'gigantic loop'
10) loops such as 'pearl necklace collar

Fig. 3. 'Deer horn'capillary.

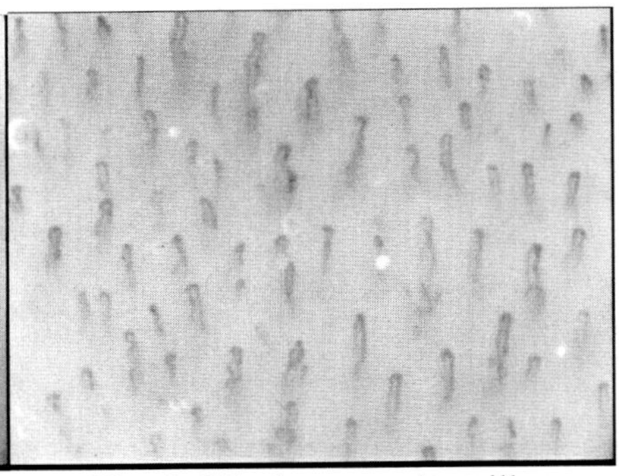

Fig. 4. 'Elephant nose' capillary.

In loop delivery we have observed:

1) ordered, regular, such as 'comb' loops
2) meandering delivery
3) focal absence
4) a gigantic loop with a few normal loops around it.

4.2. LORCA

The LORCA system is an interesting method for the study of RBC deformability and aggegability. Laser beams allow the detection of changes in RBC deformability and their aggregability [5-7].

Fig. 5. LORCA.

4.3. Oximeters

Oxygen partial pressure is measured using the Periflux 5000 (Perimed) with combi sensors.

5. RESULTS AND DISCUSSION

In patients with diabetic microangiopathy (Group B): in 50% (16/32) the capillary loop showed formations such as 'deer horns', in 72% (23/32) a formation such as 'elephant nose', and in 45% (14/32) a formation such as a 'cork screw'. In diabetics with POAD an important capillary rarefaction was found in 26% (9/32) of the patients. In glaucoma (Group C): in 84% (25/30) of the patients we observed 'capillary meandering' and images such as 'a comb'. In patients with more complicated pathology capillary rarefaction was found in 70% (21/30) of the group.

We also found an improvement in the perfusion of non-functional loops in patients who had died due to liver failure one week after liver transplantation (Group D) in 90% (5/6) of the studied cadavers. In Group E: in non-smoking hypertensives we found morphological changes such as 'tortuous loops' in 25% (6/22) of the patients, and in 47% (13/28) of the smoking hypertensive patients.

RBC deformability expressed as the elongation index (EI), and RBC aggregability expressed as $t^{1/2}$ (seconds) indicating the time to obtain peak RBC aggregability, were detected using LORCA.

We found a significant decrease in the EI of all groups compared with controls: 0.56±0.01 vs 0.59±0.01 ($p<0.04$) and also a significant decrease in $t^{1/2}$ in all groups compared with controls: 2±0.2 ($p<0.04$). The $TcpO_2$was detected at the dorsum of the right foot as a standard site representing peripheral control of microvasculature perfusion. The $TcpO_2$ appeared significantly decreased ($p<0.05$) compared with controls (Table 6).

Table 6 Data on $TcpO_2$ values in the five study groups

		TcpO2 (mmHg)	p-value
Group A	Controls	96 ± 11	*
Group B	IDDM	74 ± 9	* <0.05
	NIDDM	76 ± 8	* <0.05
Group C	Glaucoma	75 ± 9	* <0.04
Group D	Liver failure	69 ± 6	* <0.04
Group E	Smokers	70 ± 5	* <0.05
	Non-smokers	77 ± 9	* <0.03

Normal RBC deformabiliy is important in the perfusion and oxygenation of peripheral capillaries, especially in ocular tissues (retina, optic nerve, etc.). It is also possible to interfere with RBC deformability using drugs that influence the amount of intracytosolic calcium. The influence of an increase in surface RBC acetylcholinesterase (ACE) may be very important [8]. ACE is an enzyme located on the external surface of the RBC membrane. An increase in ACE induces an increase in cytosolic calcium in RBC and a subsequent decrease of RBC deformability with impairment in microcirculatory blood flow and tissue oxygenation [9-13]. We compared healthy control subjects with glaucoma patients (Group C), thereby exploring other aspects in the same glaucoma group [14-15] (Table 7).

Table 7 Data on glaucoma patients

	Type	Number of patients	Males (M)	Females (F)	Age (years)
Group A	Controls	25	15	10	36 ± 3
Group C1	Glaucoma: topical β-blockers	10	5	5	44 ± 3
Group C2	Glaucoma: carbonic anhydrase inhibitor	10	6	4	45 ± 6
Group C3	Glaucoma: prostaglandins	10	5	5	50 ± 3

6. CONCLUSIONS

This study using computerized videocapillaroscopy to investigate capillary morphology, LORCA to investigate RBC plasticity, and transcutaneous oximeter to study tissue oxygenation presents an interesting and complete methodology to explore and evaluate the microcirculation in different pathologies that induce changes in the microvasculature. These approaches allow the physician to modulate the therapy during follow-up in patients with these types of pathologies.

In addition, our data show that in glaucomatous patients topical prostaglandins and carbonic anhydrase inhibitors do not significantly (p=n.s.) alter the surface RBC ACE activity in these patients compared with controls. Topical β-blocker drugs in glaucomatous patients are able to induce a significant ($p<0.05$) increase in ACE and thus also an increase in intracytosolic calcium, finally decreasing RBC deformability and therefore tissue oxygen supply. Therefore, we conclude that carbonic anhydrase inhibitors and prostaglandin drugs do not interfere significantly with the intra-ocular microcirculation and hemorheology (particularly the optic nerve blood flow), whereas β-blockers do.

7. REFERENCES

1. Leeuwenhoek v. A. In: Clinical and experimental studies on hemorheology. Fleur C. Mokken 1996;20-21. AMC: University of Amsterdam Publisher, the Netherlands.
2. Starling EH. The Starling Law. Journal of Physiology 1895;19:312-326.
3. Piovella C. In: La capillaroscopia in Medicina. Guzzo G. 1996;1:45 EDRA Ed. Milano, Italy.
4. Bollinger J. and Fagrell B. Clinical capillaroscopy, a guide to its use in clinical research and practice. Hogrefe & Huber Publishers, Toronto, 1990.
5. Cicco G. Red blood cells (RBC) deformability, RBC aggregability, and tissue oxygenation in hypertension. Clin Hemorh Microcirc 1999;21:169-177.
6. Hardeman MR, Goedhart PT. Laser assisted optical rotational red cell analyzer (LORCA): a new instrument for measurement of various structural hemorheological parameters Clin Hemorheol 1994;14:605-618.
7. Hardeman MR, Goedhart PT Laser assisted optical rotational red cell analyzer (LORCA): Red cell aggregometry in: Hemorheology and Erythrocyte Aggregation, vol. 4 J.F. Stoltz Editions Medicales Intérnationale, Paris.
8. Ellmann GL, Courtney KD. A new rapid colorimetric determination of hemorheological acetylcholinesterase activity. Biochem Pharmacol 1961;7:88-95.
9. Cicco G. Alterazioni emoreologiche e calcio citosolico eritrocitario nell'ipertensione Arteriosa Giornale Italiano di Nefrologia 2000;17,164,c9:46.
10. Cicco G. Relationship between cytosolic calcium and red blood cell deformability during arterial hypertension. In: High Blood Pressure. ed Kurtis Milano 1999;8,1:13.
11. Cicco G, Carbonara MC, Hardeman MR. Cytosolic calcium and hemorheological patterns during arterial hypertension. Clin Hemorheol Microcirc 2001;24:25-31.
12. Cicco G. Hemorheological alterations and erythrocyte cytosolic calcium in arterial hypertension. Nephrol Dialysis Transplant 2000;15,9:A67.
13. Cicco G, Carbonara MC, Hardeman MR. Hemorheology, cytosolic calcium and carnitinaemia in subjects with hypertension and POAD. J Des Maladie Vasculaires 2000;154:20-22.
14. Zabala L, Saldanha C, Martins e Silva J, Souza-Ramalho P. Red blood cell integrity in primary open angle glaucoma: ex vivo and in vitro studies. Eye 1999;13:101-103.
15. Vetrugno M, Cicco G, Sborgia C. Red blood cell deformability, aggregability and cytosolic calcium concentration in normal tension glaucoma. Clin Hemorheol Microcirc 2004;31:295-302.

CLONIDINE ELICITS A LONG-TERM DEPRESSION IN MUCOSAL RED CELL FLUX

Artur Fournell, Olaf Picker, Ingo Schwartges, Thomas W. L. Scheeren, and Lothar A. Schwarte*

Abstract:
Objective: To evaluate the impact of clonidine on mucosal red cell flux during baseline sedation with propofol or sevoflurane, respectively.
Materials and Methods: Six healthy, chronically instrumented dogs for the measurement of cardiac output (CO) were repeatedly studied. During baseline sedation with either propofol (15 mg • kg^{-1} • h^{-1}) or sevoflurane (1.5 MAC), local tissue cell flux was assessed using laser Doppler flowmetry at the enoral mucosa. After baseline measurements, a bolus of clonidine (2.0 μg/kg) was infused within 1 min. Data are presented as mean ± SEM; Statistics: ANOVA, Scheffé's post hoc test, $p < 0.05$.
Results: Clonidine significantly reduced CO from 75 ± 4 and 75 ± 6 ml • kg^{-1} • min^{-1} (sedation with propofol or sevoflurane, respectively) to 40 ± 3 and 49 ± 5 ml • kg^{-1}• min^{-1}, however, with almost complete recovery to baseline after 30 min (70 ± 4 and 71 ± 6 ml • kg^{-1} • min^{-1}, NS from baseline). Similarly, clonidine decreased mucosal red cell flux by 44 ± 8% and 54 ± 4%. However, mucosal perfusion did not return to baseline (-25 ± 5% and -27 ± 3%).
Conclusions: In spite of the rapid return to baseline in systemic perfusion, the mucosal red cell flux of the enoral mucosa remained markedly reduced after a single bolus of clonidine. Given the crucial role of preserved microcirculatory perfusion for an intact mucosal barrier function, our data suggest that clonidine might impair this important mechanism to prevent the translocation of bacteria and endotoxins into the systemic circulation.

1. INTRODUCTION

The pharmacodynamic characteristics of α_2-agonists show some aspects that recommend these drugs for the application in critically ill patients. The site for their sedative action is

* Department of Anesthesiology, Heinrich-Heine-University Duesseldorf, 40001 Duesseldorf, Germany, E-mail fournell@uni-duesseldorf.de

located in the locus coeruleus of the brain stem, whereas the principal site for the analgesic action is probably in the spinal cord. In the heart, the most important action of α_2-agonists is a decrease in tachycardia (through a blockade of the cardioaccelerator nerve) and bradycardia (through a vagomimetic action). In the peripheral vasculature, they exhibit both a vasodilatory action *via* sympatholysis and vasoconstriction mediated through receptors in the smooth muscle cells. Moreover, α_2-agonists have antishivering properties.[1]

However, in critically ill patients it is of utmost importance to not only consider potential beneficial effects of any drugs but to also be aware of possible side effects on tissue perfusion. This applies especially to the splanchnic region, a site usually referred to as the motor of multiple organ failure.[2] Herein, microvascular blood flow disturbances have been identified as a key factor.[3] Therefore, it is of pertinent importance that any medication does not further compromise splanchnic perfusion.

Although monitoring tools to assess mucosal blood flow in the splanchnic region are available, e.g., laser Doppler flowmetry, their application in surgical patients is often restricted due to technical problems, which may interfere with the measuring process in the perioperative period, e.g., acute gastrointestinal bleeding. To avoid these problems, the enoral mucosa has been advocated as an appropriate alternative monitoring site.[3, 4]

Therefore, we sought to evaluate the impact of clonidine on enoral mucosal perfusion. Furthermore, since the effects of this drug might be modulated by the kind of baseline sedation, we studied these effects during sedation with propofol and sevoflurane, respectively.

2. MATERIALS AND METHODS

2.1. Animals

The data were derived from repetitive experiments on healthy dogs (foxhounds of both sexes, n = 6 dogs, body weight 25 - 35 kg) treated in accordance with the NIH guidelines for animal care and with approval of the local District Governmental Animal Investigation Committee.

2.2. Animal Care and Monitoring

The dogs were studied twice, i.e., during baseline sedation with either propofol ($15\ mg \cdot kg^{-1} \cdot h^{-1}$) or sevoflurane (1.5 MAC) on different days with mechanical ventilation (tidal volume 12 ml/kg). To preclude the influence of varying arterial partial pressures of carbon dioxide on local tissue blood flow[5] normocapnia was maintained by adjusting the respiratory rate (~20/min) and verified by continuous capnography (end-tidal CO_2 35 mmHg) and intermittent blood gas analysis. The depth of sedation was controlled using the bispectral index method (Model A-2000, Aspect Medical Systems, Newton, MA, target value: 50 – 70). During the experiments, the dogs were lying on their right side, covered with warming blankets to maintain body temperature within the physiological range for dogs (37.0 - 38.5 °C, rectal thermoprobe).

2.3. Measurements

Monitoring consisted of heart rate (ECG-triggered cardiotachometer) and mean arterial (aortic) pressure (MAP, Gould-Statham pressure transducers P23ID, Elk Grove, IL). For the continuous measurement of cardiac output (CO) transit-time flowmeters (20 mm S-series, Transonic, Ithaca, NY) were implanted chronically around the pulmonary artery at least four weeks prior to the experiments and calibrated as previously described.[6] A laser Doppler flowmeter (MBF3D, Moor Instruments, Axminster, Devon, UK) was used to provide non-invasive, real-time measurements of local tissue red cell flux (LDF) at the enoral mucosa. All data were stored on personal computer systems after analog to digital conversion for later analysis.

2.4. Study Protocol

After assessing baseline values during sedation with either propofol or sevoflurane a single bolus of clonidine (2.0 μg/kg) was given within 1 min. Repeated measurements were made after 2 min (to observe the maximum effect) and after 30 min, a time sufficient to allow the systemic variables (CO and MAP) to return to baseline values.

2.5. Statistical Analysis

Statistical calculations and analysis were performed using StatView (version 5.0, SAS Institute Inc., Cary, NC). The results obtained are given as mean ± SEM. One-way analysis of variance (ANOVA) for repeated measurements was applied to analyze continuous variables measured over time. In case of significance Scheffé's post hoc test was applied to identify significant differences. Statistical significance was assumed if the α error was smaller than 5%.

3. RESULTS

The type of baseline sedation did not influence the effects of clonidine on the variables of systemic perfusion. Similarly, during sedation with both propofol and sevoflurane, the bolus injection of clonidine reduced cardiac output and increased mean arterial pressure. Although the impact of clonidine on CO and MAP showed some differences for the two sedatives, they did not reach the level of significance. In detail, CO decreased from 75 ± 4 and 75 ± 6 $ml \cdot kg^{-1} \cdot min^{-1}$ to 40 ± 3 ($p < 0.0001$) and 49 ± 5 $ml \cdot kg^{-1} \cdot min^{-1}$ ($p < 0.0001$) during sedation with propofol or sevoflurane, respectively (Fig. 1). MAP increased from 83 ± 7 and 61 ± 1 mmHg to 100 ± 8 ($p < 0.05$) and 87 ± 3 mmHg ($p < 0.0001$) during sedation with propofol or sevoflurane, respectively. Both variables of systemic perfusion returned almost completely to baseline values within 30 min (CO: 70 ± 4 and 71 ± 6 $ml \cdot kg^{-1} \cdot min^{-1}$; MAP: 85 ± 3 and 61 ± 2 mmHg for propofol or sevoflurane, respectively, non significant from baseline).

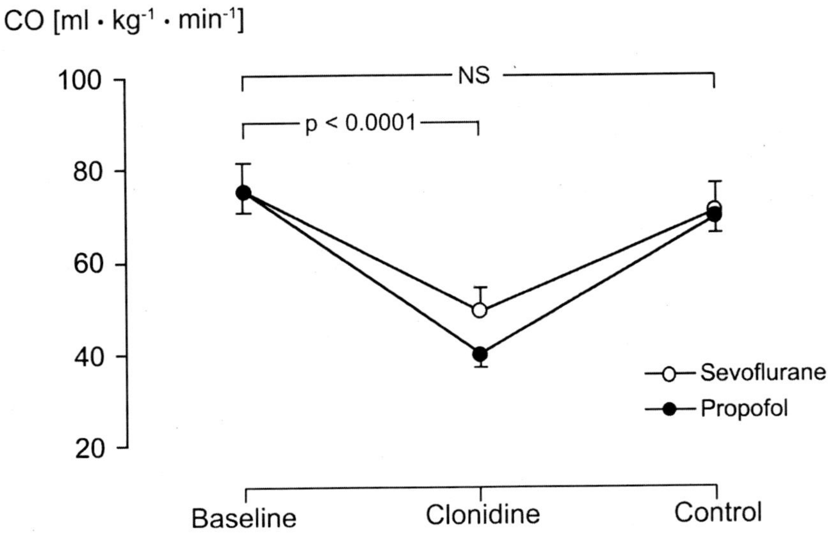

Figure 1. The infusion of clonidine transiently decreased cardiac output (CO) with almost full recovery after 30 min. "Clonidine" refers to measurements taken 2 min after the injection of clonidine, a time sufficient to observe the maximal effects of the α_2-agonist; "Control" refers to repeated measurements 30 min after the injection of clonidine, a time sufficient to allow the systemic variables to return to baseline values. Data (during sedation with either propofol or sevoflurane, respectively) are given as mean ± SEM for n = 6 dogs.

In accordance with the diminution in CO clonidine significantly reduced the local tissue red cell flux of the enoral mucosa (Fig. 2). Although these effects were slightly more pronounced during sevoflurane compared to propofol, the difference did not reach the level of significance. LDF decreased by 44 ± 8% ($p < 0.0001$) from baseline during sedation with propofol and by 53 ± 4% ($p < 0.0001$) during sedation with sevoflurane, respectively. However, even though LDF increased significantly after 30 min it remained significantly below baseline values (25 ± 5% for propofol, $p < 0.0001$ and 26 ± 3% for sevoflurane, $p < 0.0001$, respectively).

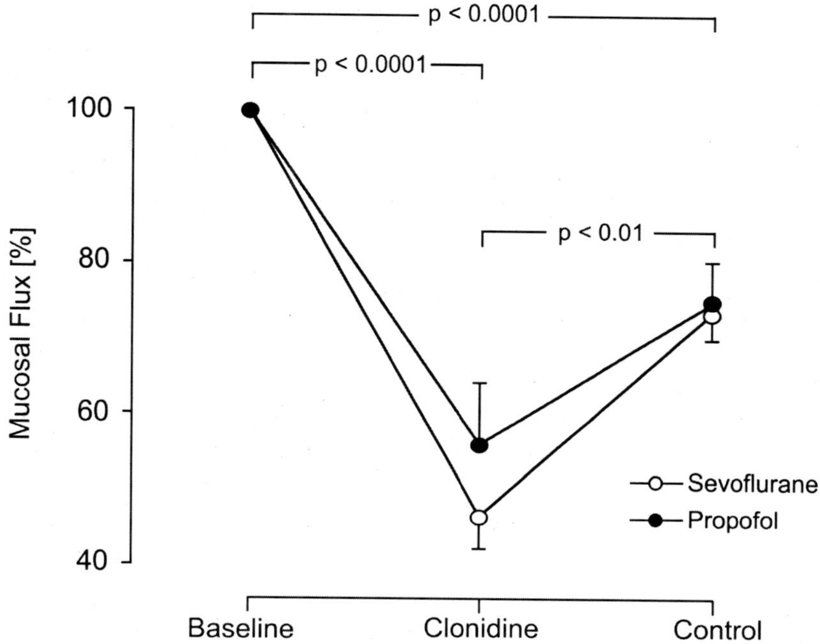

Figure 2. The infusion of clonidine induced a sharp and sustained decline in the red cell flux of enoral mucosa (LDF). "Clonidine" refers to measurements taken 2 min after the injection of clonidine, a time sufficient to observe the maximal effects of the α_2-agonist; "Control" refers to repeated measurements taken 30 min after the injection of clonidine, a time sufficient to allow the systemic variables to return to baseline values. Data (percentage of the baseline values during sedation with either propofol or sevoflurane, respectively) are given as mean ± SEM for n = 6 dogs.

4. DISCUSSION

This study was designed to evaluate the impact of clonidine on systemic hemodynamics and local tissue red cell flux. The main result is that clonidine, irrespective of the type of baseline sedation, elicited a long-term depression in the red cell flux of the enoral mucosa despite already normalized systemic variables of tissue perfusion.

These findings confirm earlier results by our group that the effect of any interventions on regional perfusion cannot be derived necessarily by monitoring systemic variables,[7] i.e., we have to access the regional circulation to be able to observe reliably the influence of any therapeutic action on regional perfusion. This issue has to be stressed, especially in view of the heterogeneous action of adrenergic agonists and antagonists due to the different expression of their respective receptors in various tissues.[8]

We are aware that experimental data may not be translated directly to the clinical setting. Furthermore, we hypothesized that the enoral mucosa can serve as an appropriate monitoring site to reflect concomitant changes in splanchnic mucosa. However, in our study, the attenuation in LDF persisted despite normalized systemic hemodynamics. Thus the observed microvascular blood flow alterations in the mucosal layer and hence the likelihood of an evolving mucosal ischemia could disturb the mucosal barrier function, thereby initiating the translocation of bacteria and endotoxins into the systemic circulation.[9]

5. CONCLUSIONS

Our data show that even a short-term application of clonidine induces a prolonged reduction in the red cell flux of the enoral mucosa. As these findings cannot be derived from variables of systemic hemodynamics the regional circulation has to be monitored to enable us to reliably detect the impact of vasoactive drugs on local tissue perfusion. Furthermore, the observed reduction in mucosal red cell flux might indicate a disturbance of the mucosal barrier function.

6. REFERENCES

1. T. Kamibayashi and M. Maze, Clinical uses of α_2-adrenergic agonists, *Anesthesiology* **93**, 1345-1349 (2000).
2. C. J. Carrico, J. L. Meakins, J. C. Marshall, D. Fry, and R. V. Maier, Multiple-organ-failure syndrome, *Arch. Surg.* **121**, 196-208 (1986)
3. D. De Backer, J. Creteur, J. C. Preiser, M. J. Dubois, and J. L. Vincent, Microvascular blood flow is altered in patients with sepsis, *Am. J. Respir. Crit. Care. Med.* **166**, 98-104 (2002).
4. P. E. Marik, Sublingual capnography: a clinical validation study, *Chest* **120**, 923–927 (2001).
5. L. A. Schwarte, I. Schwartges, A. Fournell, T. W. L. Scheeren, and O. Picker, Hypercapnia increases gastric mucosal oxygenation during hemorrhagic shock, *Intensive Care Med.* 31(Suppl. 1), S578 (2005).
6. O. Picker, A. Schindler, and T. W. L. Scheeren, Accuracy and reproducibility of long-term implanted transit-time flow probes in dogs, *Intensive Care Med.* **26**, 601-607 (2000).
7. A. Fournell, T. W. L. Scheeren, and L. A. Schwarte, PEEP decreases oxygenation of the intestinal mucosa despite normalization of cardiac output, *Adv. Exp. Med. Biol.* **454**, 435-440 (1998).
8. T. W. L. Scheeren, L. A. Schwarte, S. A. Loer, O. Picker, and A. Fournell, Dopexamine but not dopamine increases gastric mucosal oxygenation during mechanical ventilation in dogs, *Crit. Care. Med.* **32**, 881-887 (2002).
9. S. M. Pastores, D. P. Katz, and V. Kvetan, Splanchnic ischemia and gut mucosal injury in sepsis and the multiple organ dysfunction syndrome, *Am. J. Gastroenterol.* **91**, 1697-1710 (1996).

REAL-TIME, AUTOMATED, FLUOROPHORE MEDIATED MULTI-CARDIAC MARKER BIOSENSING SYSTEM WITH NANO-METALLIC PARTICLE REAGENT

Bin Hong, Liang Tang, Yongjie Ren, and Kyung A. Kang*

Abstract: Cardiovascular disease (CVD) is the leading cause of death in US. Early and accurate diagnosis of CVD is crucial to save many lives, especially for the patients suffering the heart attack. Accurate and fast quantification of cardiac muscle specific biomarkers in the blood enables accurate diagnosis and prognosis and timely treatment of the patients. A prototype of fiber-optic, multi-analyte, immuno-biosensing system integrated with an automatic flow control unit has been in development to quantify four important cardiac markers in blood plasma accurately, rapidly and simultaneously. The validity of the sensor was, however, challenged because the concentrations of two markers are only at tens of picomolar level. Here, plasmon rich nano-metallic particles and selected biocompatible solvents were developed for fluorescence enhancement to improve the sensitivity of our fluorophore mediated biosensing. By applying the nano-metal particle and the solvent, the sensitivity of single cardiac marker sensors were increased by 1.5 ~ 3 times. By using the fluorescence enhancing nano-metallic particle reagents, simultaneous quantification of four cardiac markers in plasma is currently possible, using 3-cm-sensor within 10 minutes at an average signal-to-noise ratio of 20.

1. INTRODUCTION

Heart attack, a leading threat for human life, causes 250,000 deaths each year in the United States.[1] During the heart attack, some biomolecules are released from the damaged heart muscle to blood. Therefore, diagnosis and prognosis can be performed by quantifying these cardiac muscle specific biomolecules (cardiac markers) in blood. An automated, fiber optic, fluorophore mediated, immuno-sensing system is being developed

* Bin Hong, Liang Tang, Yongjie Ren, and Kyung A. Kang, Department of Chemical Engineering, University of Louisville, Louisville, KY 40292.

in our research group, to simultaneously quantify the four most important cardiac markers in real-time. These four markers are myoglobin (MG), C-reactive protein (CRP), cardiac Troponin I (cTnI), and B-type natriuretic peptide (BNP). Clinically important sensing ranges that we set for cTnI and BNP in blood plasma are 31 ~ 310 pM and 26 ~ 260 pM[2-3], respectively. These ranges are very low, challenging the sensor development.

Nano-sized metallic particles (NMPs) with high plasmon density have unique properties for fluorescence enhancement.[4-6] When these NMPs are placed at an appropriate distance from a fluorophore, they can attract the free electrons of the fluorophore that are normally used in self-fluorescence quenching to their surface plasmon field and, hence, enhance the fluorescence. The efficiency of the electron transfer has been verified to be dependent on the separation between a NMP and a fluorophore, the NMP size, and the quantum yield of a fluorophore. As an effective method to control the distance between these two entities, Hong and Kang[4-6] have immobilized self-assembled monolayers (SAMs) at various thicknesses on the surface of NMPs. Among SAM linked nanogold particles (NGPs) tested, 5 nm NGPs linked with a 2 nm thick SAM (5nmNGP-SAM2nm) were found to be very effective fluorescence enhancers.

Some bio-compatible organic solvents were also found to enhance fluorescence significantly. When the NGP was combined with the solvent (Nanogold Particle Reagent; NGPR), the enhancement level was the sum of the portion by the NGP and that by the solvent.

In this paper, the effects of various enhancers, i.e., NMPs (nanogold particle and nanosilver particle), solvents, and their mixture, defined as nanometal particle reagents (NMPRs), on fluorophore mediated biosensing are reported. Results of simultaneous four-cardiac marker sensing using these fluorescence enhancers are also presented.

2. MATERIALS, INSTRUMENTS AND METHODS

2.1. NMPs and solvents

Nanogold particles (size: 5 and 20 nm) and nanosilver particles (size: 20 nm), coated with tannic acid (approximately 3 nm), were obtained from Ted Pella (Redding, CA). 16-mercaptohexadecanoic acid [approximately 2 nm; Sigma/Aldrich (St. Louis, MO)] was used for SAM preparation. Hexane, 2-propanol, 1-butanol, tetrahydrofuran (THF), methanol, ethanol and dimethylsulfoxide (DMSO) were tested as solvents (Sigma/Aldrich).

2.2. Cardiac marker sensors, assay protocol, and fluorophores

BNP (Bachem; Torrance, CA), and two different anti-BNP monoclonal antibodies (Strategic Biosolutions; Newark, DE), cTnI, MG, CRP and their respective murine monoclonal antibodies (Fitzgerald; Concord, MA) were used to prepare BNP, cTnI, MG, and CRP sensors. The fluorophores, Fluorolink™ Cyanine 5 (Cy5; 649/670 nm for excitation/emission) and Alexa Fluor® 647 (AF647; 649/666 nm for excitation/emission), were purchased from Amersham Pharmacia Biotech (Uppsala, Sweden) and Molecular Probes (Eugene, OR), respectively.

Quartz fibers used for sensors and the fluorometer (Analyte 2000) were purchased from Research International (Monroe, WA). The cardiac marker sensors were prepared following the protocols established by Spiker and Kang.[7] The sensors perform a fluoro-immunoassay on the sensor surface: The first monoclonal antibodies (1° Mab) against the marker are immobilized on the sensor surface. The target markers specifically bind to 1° Mab and then the second monoclonal antibodies tagged with a fluorophore (Fluorophore-2°Mab) bind to the markers, forming a sandwich protein complex. When the bound fluorophores are excited, the emission fluorescence is correlated with the concentration of the target marker in samples. The methods used to prepare NMPs and NMPRs, and the protocols for fluorophore mediated biosensing with and without enhancers were described by Hong and Kang.[4-5] In sensing with various enhancers, the baseline was taken with the enhancer before the target analytes bind to the surface, and new steps were added after the immuno-reaction between the analyte and the fluorophore-2°Mab. The sensor was washed, and then the enhancer was added and fluorescence measurement was performed.

3. RESULTS AND DISCUSSION

Our bioensing system is currently being developed as an automatic device for bedside, real-time heart attack diagnosis and prognosis. In the system, four single-cardiac marker sensors were connected in series, to quantify four cardiac markers simultaneously. The sample and the reagents circulate at an optimal velocity in the unit for effective mass transfer. The entire system is automatically controlled by a computer program for user-friendliness and high consistency in the sensing performance. However, most importantly, the sensitivity of the BNP and cTnI sensors needed to be improved in the range of our interest, i.e. tens of picomolar level. The fluorescence enhancer was, therefore, directly related to the success of our cardiac marker sensing system.

3.1. Cardiac marker biosensing using nano-metal particles (NMPs)

Hong and Kang[4-6] have previously proven that nanogold particles can enhance the fluorescence *via* their strong plasmon field, for both free fluorophores and fluorophore mediated biosensing. Nanosilver particles (NSP) also have a strong plasmon field on the particle surface.[8] Therefore, 3 nm SAMs linked 20 nm NSP, the smallest, commercially available one, was tested and the enhancement level was compared with that by the NGP at the same size (Fig. 1). For this study, a 1.5 cm BNP sensor was used. Without enhancers, the sensor provided very low signals (< 30 pA; Fig. 1, ■). By using the NSP, 30% ~ 35% enhancement was obtained (Fig. 1, ◇). NGP at the same size with same surfactant was also tested. 81% ~ 138% enhancement was shown (Fig. 1, ◆), higher than that of NSP. Hence, NGP was found to be more effective for the fluorescence enhancement than NSP, probably due to its higher electron affinity (2.3086 eV) than that of the NSP (1.3030 eV).

3.2. Cardiac marker biosensing using the solvents

Some organic solvents were also found to enhance the fluorescence by several fold, while not affecting the sensor reusability.[4-6] The mechanisms of the enhancement by solvent were by the molecular interaction between the fluorophore and the solvent, and by the conformational changes in protein complex in a solvent. From our study results, it was found that solvents often affect the *Trans-Cis* isomerization of the double bonds in the fluorophore, shift the fluorophore emission spectrum to "red", and also shrink the fluorophore linked protein complex on sensor surface (data not shown).

To select the optimal solvent for biosensing, various solvents, i.e., hexane, 2-propanol, 1-butanol, THF, methanol, ethanol, DMSO, and PBS buffer, were tested with the BNP sensor (data not shown). Among the solvents, 1-butanol presented the highest enhancement i.e., showing a signal of 8 times that of PBS buffer. Figure 2 shows the performance of a BNP sensor using 1-butanol and ethanol as enhancers. By 1-butanol, the sensitivity of the BNP sensor increased by 5 ~ 7 times, 4 ~ 6 times higher than that of ethanol. Also, neither of these solvents negatively affected the reusability of the sensors.

3.3. Cardiac marker biosensing using nano-metal particle reagents (NMPRs)

To maximize the signal enhancement, the NMP and the solvent were combined to form the nano-metal particle reagents (NMPRs) and these reagents were tested for the sensitivity improvement in multi-cardiac marker sensing. 5nmNGP-SAM2nm in ethanol, an effective NGPR previously tested for the protein C (PC) sensor, was tested. Figure 3 shows that the NGPR can improve the sensitivity of all four individual cardiac marker sensors (by 1.5 ~ 3 times). Also, the enhancement level for the cardiac marker sensors by NGPR was even greater than the sum of the individual effects by the NGP and by the solvent (data not shown).

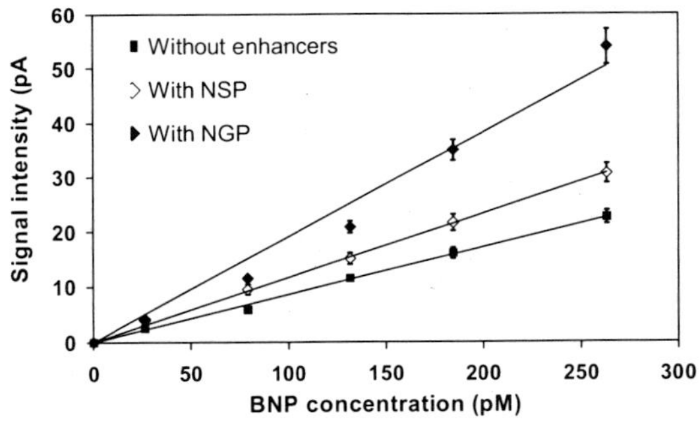

Figure 1. Fluorescence enhancement in BNP biosensing by 20nmNSP-SAM3nm and 20nmNGP-SAM3nm. [Experimental conditions: sensor size, 1.5 cm; sample/AF647-2°Mab incubation time, 3/4 min, respectively; flow velocity, 1.2 cm/sec].

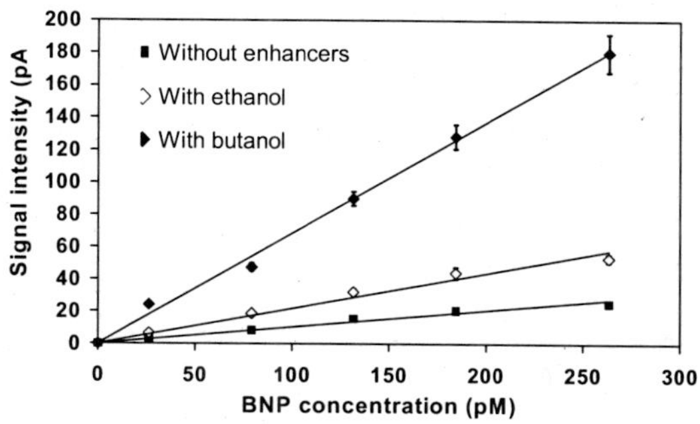

Figure 2. Fluorescence enhancement in BNP biosensing by ethanol and 1-butanol. [Experimental conditions: sensor size, 1.5 cm; sample/AF647-2°Mab incubation time, 3/4 min; flow velocity, 1.2 cm/sec].

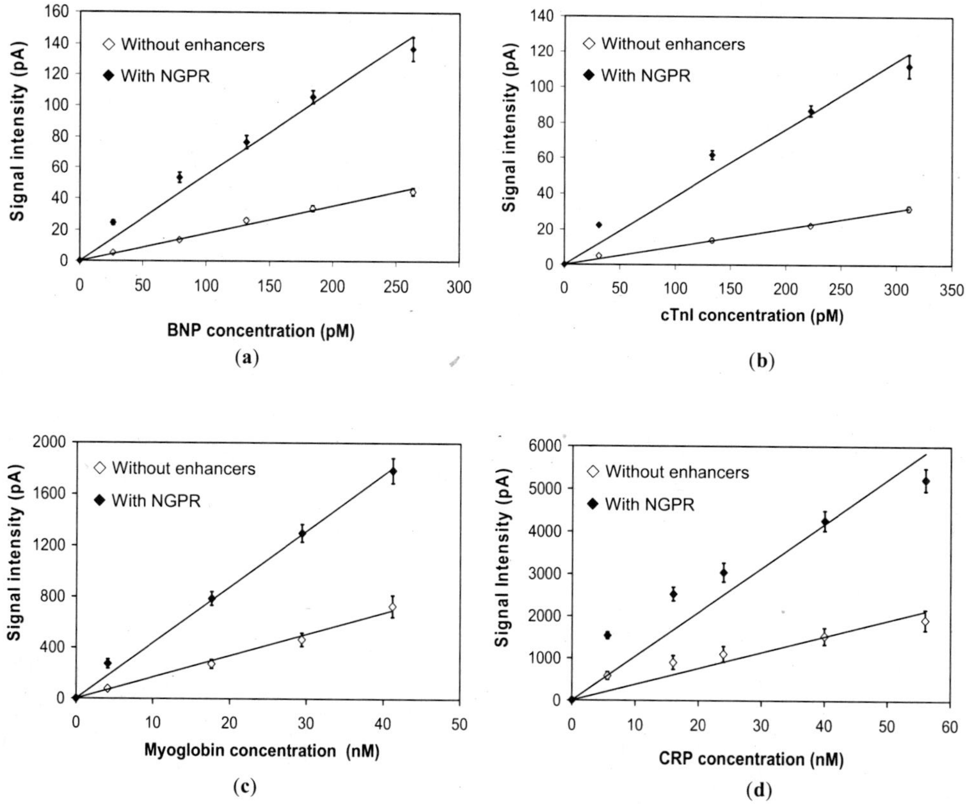

Figure 3. NGPR effect on the biosensing of (a) BNP, (b) cTnI, (c) MG, and (d) CRP. [Experimental conditions: sensor size, 3 cm; sample/AF647-2°Mab incubation time, 3/4 min; flow velocity, 1.2 cm/sec; NGPR: 5nmNGP-SAM2nm in ethanol].

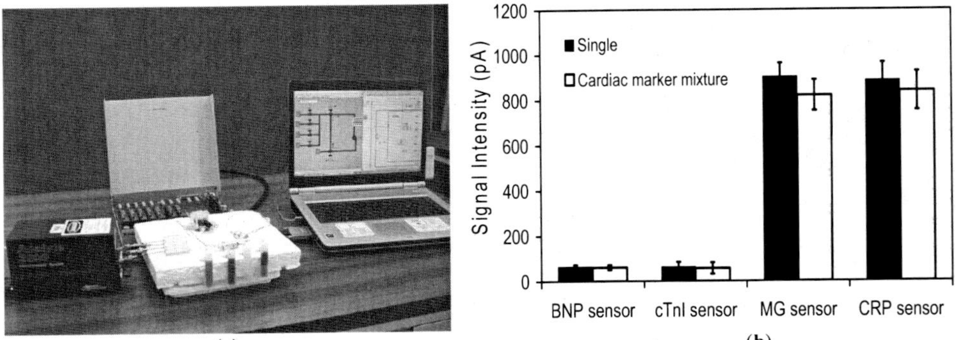

(a) (b)

Figure 4. Automatic, four-analyte sensing system: (a) System components, and (b) Comparison between the individual sensing and multi-sensing [Experimental conditions: sensor size, 3 cm sensors; 184 pM BNP; 222 pM cTnI; 41 nM MG; 24 nM CRP; sample/AF647-2°Mab incubation time, 3/4 min; flow velocity, 1.2 cm/sec; enhancer, ethanol].

3.4. Automatic, simultaneous cardiac marker biosensing system

As previously stated, an automatic biosensing system is being developed by our research group for simultaneous quantification of the four cardiac markers. Figure 4a shows the entire sensing system: from left to right, the fluorometer (Analyte 2000), the sensing unit with flow control device, and the computer control system. The assay protocol was programmed and the entire assay was run by clicking the computer. All four sensors detect the fluorescence by a single fluorophore and quantify their amounts in the sample. Figure 4b shows the comparison of the signal intensity of the individual sensors (white bars) and the multi-sensing system (black bars) at a medium biomarker concentration of each range. The signal intensities of single sensors were slightly higher than those of the multi-sensor, possibly due to the mixing of the samples in the dead volume along the tubing in the multi-sensing system. This issue is expected be minimized as this prototype is translated to a multi-sensing biochip. This preliminary result demonstrated that our automatic, cardiac marker sensing system is capable of simultaneously quantifying four cardiac markers within 10 minutes, at an average S/N ratio of 20.

4. CONCLUSIONS

Nano-metal particles (e.g. NGP and NSP) demonstrated to significantly enhance the sensitivity of the fluorophore mediated biosensing. NGP exhibits higher enhancement than NSP possibly due to its higher electron affinity. Organic solvents can also enhance the fluorescence significantly and as much as 7 times fluorescence enhancement was obtained by 1-butanol in BNP sensing. NMPRs, the mixture of NMP and the biocompatible solvent, showed the fluorescence enhancement by the NMP and by the solvent additively. A prototype of an automated four-cardiac marker sensing system was developed to simultaneously quantify four markers with the help of NGPR. The preliminary result showed that the signals obtained from this system were comparable to those using a single sensor, enabling four cardiac marker quantification within 10 minutes at an average S/N ratio of 20.

5. ACKNOWLEDGEMENTS

The authors thank the Kentucky Science and Engineering Foundation (KSEF-148-502-03-55) for financial support for the metallic nanoparticle study, and the National Science Foundation (BES-0330075) for the development of the cardiac marker biosensing system.

6. REFERENCES

1. American Heart Association. Heart Disease and Stroke Statistics, Update, 10-12 (2005).
2. M. S. Sabatine , D. A. Morrow, J. A. de Lemos, C. M. Gibson, S. A. Murphy, N. Rifai, C. McCabe, E. M. Antman, C. P. Cannon, and E. Braunwald, Multimarker approach to risk stratification in non-ST elevation acute coronary syndromes: Simultaneous assessment of troponin I, c-reactive protein, and b-type natriuretic peptide, *Circ.* **105**, 1760-1763 (2002).
3. A. S. Maisel, P. Krishnaswamy, R. M. Nowak, J. McCord, J. E. Hollander, P. Duc, T. Omland, A. B. Storrow, W. T. Abraham, A. H. B. Wu, P. Clopton, P. G. Steg, A. Westheim, C. W. Knudsen, A. Perez, R. Kazanegra, H. C. Herrmann, and P. A. McCullough, Rapid measurement of B-type natriuretic peptide in the emergency diagnosis of heart failure, *New Engl. J. Med.* **347**, 161-167 (2002).
4. B. Hong and K. A. Kang, Fluorescence enhancers for fluorophore mediated biosensors for cardiovascular disease diagnosis, *Advances in Experimental Medicine and Biology* **578**, in press (2006).
5. B. Hong and K. A. Kang, Biocompatible, nanogold-particle fluorescence enhancer for fluorophore mediated, optical immunosensor, *Biosensors Bioelectronic.* **21**(7), 1333-1338 (2006).
6. K. A. Kang and B. Hong, Biocompatible nano-metal particle fluorescence enhancers, *Critical Reviews in Eukaryotic Gene Expression* **16**(1), 45-60 (2006).
7. J. O. Spiker, K. A. Kang, W. N. Drohan, and D. F. Bruley, Preliminary Study of Biosensor Optimization for the Detection of Protein C, *Advances in Experimental Medicine and Biology* **454**, 681-688 (1998).
8. J. R. Lakowicz, J. Malicka, S. D'Auria, I. Gryczynski, Release of the self-quenching of fluorescence near silver metallic surfaces. *Anal. Biochem.* **320**, 13–20 (2003).

PTEN AND NDUFB8 ABERRATIONS IN CERVICAL CANCER TISSUE

S.M. Hsieh*, D. J. Maguire *, N. A. Lintell *, M. McCabe, *
& L.R. Griffiths# [1]

Abstract: Cervical cancer is one of the world's major health issues. Despite many studies in this field, the carcinogenetic events of malignant conversion in cervical tumours have not been significantly characterised. The first aim of this project was to investigate the mutation status of the tumour suppressor gene- Phosphatase and Tension Homolog (PTEN)- in cervical cancer tissue. The second aim of this study was the analysis in the same cervical cancer tissue for aberrations in the mitochondrial electron transport chain subunit gene NDUFB8, which is localised to the same chromosomal contig as PTEN. The third aim was the evaluation of the potential therapeutic anti-cancer drug 2,4-Thiazolidinediones (TZDs) and its affect in regulating the PTEN protein in a cervical cancer cell line (HeLa).

To approach the aims, paraffin-embedded cancerous cervical tissue and non-cancerous cervical tissue were obtained. DNA recovered from those tissues was then used to investigate the putative genomic changes regarding the NDUFB8 gene utilising SYBR Green I Real-Time PCR. The PTEN gene was studied via Dual-Labelled probe Real-Time PCR. To investigate the protein expression change of the PTEN protein, HeLa cells were firstly treated with different concentrations of 2,4-Thiazolidinediones and the level of PTEN protein expression was then observed utilising standard protein assays.

Results indicated that there were putative copy-number changes between the cancerous cervical tissue and non-cancerous cervical tissue, with regard to the PTEN locus. This implies a potential gain of the PTEN gene in cancerous cervical tissue. With regards to normal cervical tissue versus cancerous cervical tissue no significant melting temperature differences were observed with the SYBR Green I Real-Time PCR in respect to the NDUFB8 gene. A putative up-regulation of PTEN protein was observed in TZD treated HeLa cells.

1
*School of Biomolecular and Biomedical Science and # School of Medical Science, Griffith University, Queensland 4111, Australia.

1. INTRODUCTION

Cervical cancer is one of the world's major health issues. The carcinogenetic events of malignant conversion in cervical tumours have not been significantly characterised [1]. Phosphatase and Tension Homolog (PTEN) is a tumour suppressor gene located at chromosome 10q23. PTEN participates in the PI3 pathway, which is linked to p53 regulation and apoptosis. Mutations/deletions of the PTEN gene have been described in various human malignant tumours, including endometrial, ovarian, prostate, breast, thyroid, head and neck, renal, small cell lung cancer, primary melanoma, glioblastoma, lymphoma, hepatocellular carcinoma and cervical cancer [2,3].

Mitochondria are the extra-nuclear cellular organelle that contains its own DNA. These genes encode several subunits of the Electron Transport Chain (ETC) [4]. Mitochondria are responsible for the powering and the maintenance, i.e. the initiation and regulation of apoptosis, of the cell [4]. It is the malfunction of the apoptosis pathway that has been fundamentally linked to the initiation and progression of carcinogenesis [4]. Differences in cellular metabolism between normal and neoplastic tissue that some cell types decrease oxidative phosphorylation and increase glycolytic energy production during tumourigenesis has been proposed [4]. In addition, physiological changes of mitochondria in cancer cells have been reported [4]. A recent publication from our lab indicated that DNA aberrations in nuclear encoded subunits of the ETC might play a role in the development of the precancerous SK lesion. SK is a neo-plasm that shares squamous epithelial derivations and progression characteristics with cervical cancer [5].

CGH analysis of NMSC by Ashton et al. (2001) indicated that gross regions of certain chromosomes were of an anomalous nature, and contained genes associated with the development of NMSC as well as nuclear-encoded genes of the ETC [6].

Using this observation, this study analysed the nuclear-encoded ETC subunit gene NDUFB8, which is localised to the same chromosomal contig region as the PTEN, for aberrations in DNA derived from archival cervical cancer samples.

The Thiazolidinedione (TZD) class of Peroxisome Proliferator-Activated Receptor (PPAR)γ ligands are known for their anti-type two diabetic, anti-inflammatory and anti-cancer effects [7]. TZDs effectively suppress the growth of several human cancer cell lines including colon, breast, prostate, non-small cell lung cancer, thyroid, gastric, hypercellular, pancreatic liposarcoma cancer cell lines [7]. It is assumed that activation of PPARγ mediates TZDs anti-cancer activity [7]. PPARγ receptor is a ligand activated nuclear receptor and has been shown to regulate differentiation and/or cell growth in a number of cell types. It has been shown that PPARγ is able to bind two response elements in the genomic sequence upstream of the PTEN gene. Studies have proposed that PPARγ has a direct effect on the PTEN gene [8].

2. METHODOLOGY

To analyse the selected genes, paraffin-embedded cancerous and non-cancerous cervical tissue were obtained from Dr Peter Dash, Mater Hospital, Brisbane, Australia. The tissues were deparaffined using standard xylene-deparaffin procedure. The DNA was recovered using the standard TRIzol method. DNA was isolated using the QIAamp DNA Blood Mini Kit (Qiagen PTY Ltd, Doncaster, Victoria) and following the manufacture's instruction.

Indicated in Table 1 are the probe and primer sequences of the PTEN, NDUFB8 and Tubulin-α-8 (housekeeping) genes. The NDUFB8 gene was analysed utilising SYBR Green I Real-Time PCR and the PTEN gene was studied via Dual-Labelled probe Real-Time PCR.

Table 1. Probe/Primer sequences, $MgCl_2$ concentrations, and melt temperatures.

NDUFB8 (Primer)	Forward	TAC CCT CTC TGT CTC ACC	2.5uM, 60°C
NDUFB8 (Primer)	Reverse	CGC TTC TGA CCT GTT CTC	2.5uM, 60°C
PTEN (Primer)	Forward	CCC AGT CAG AGG CGC TAT GT	2.5uM, 60°C
PTEN (Primer)	Reverse	GAA ACA ACA GTG CCA CTG GTC	2.5uM, 60°C
PTEN (Probe)		TCC AGA TGA TTC TTT AAC AGG TAG CTA TAA T	2.5uM, 60°C
Tubulin α 8 (Primer)	Forward	GCA GCC AAC AAC TAT GCC CG	2.5uM, 60°C
Tubulin α 8 (Primer)	Reverse	CTT CCG TAT GCG GTC CAG CA	2.5uM, 60°C
Tubulin α 8 (Probe)		TCA ATG CTC TCC TTG CCC ACC GTG TAG T	2.5uM, 60°C

The SYBR Green I and the Dual-Labelled Real-Time PCR reactions both comprised 2μl DNA and 18μl of Master Mix. The composition of the Master Mix is detailed in [6].

To investigate the protein expression change of the PTEN protein, HeLa cells, obtained from Dr Ann McDonnell, Griffith University, Brisbane, Australia, were cultured in sterile RPMI-1640 medium (Sigma-Aldrich) supplemented with 5000μl/ml Penicillin/Streptomycin, sodium bicarbonate and 10% fetal calf serum (FCS). All cells were grown in sterile 25cm^2 tissue culture flasks (Becton Dickinson and Company, Franklin Lakes, NJ, USA) and incubated in a 37°C incubator. The cells were passaged every one to two days.

The cultured HeLa cells were treated with different concentrations of 2,4-Thiazolidinediones (TZD). $C_3H_3NO_2S$ (Sigma-Aldrich) was dissolved in Dimethylsulphoxide (DMSO) to make a 1 Molar TZD stock solution (MW: 117.13, 2.928g of TZD in 15 ml of sterile PBS and 10 ml of DMSO). This experiment investigated on the following concentrations: 0.125μM, 0.25μM, 0.5μM, 1μM and 2μM of TZD solution. For each semi-confluent culture flask, 100μl of medium was discarded and 100μl of filter-sterilized 2,4-Thiazolidinedione solution was added. The flasks were incubated for 24 hrs at 37°C and washed with sterile PBS. Trypsinised cells were collected by centrifugation at 12,000 RPM for 15 min. The collected cell pellets were washed with sterile PBS. Cell pellets were resuspended in 80μl of lysis buffer (0.125M Tris-HCl, pH 6.8 containing 5% SDS), and heated to 100°C for five min with mechanical shearing by pipetting the suspension up and down in a Hamilton syringe. The supernatants were recovered by centrifuging the cell suspension at 12,000 RPM for 20 min at 4°C. The protein content of the cell extracts was estimated using the DC Protein Assay. The level of PTEN protein expression was then observed utilising standard protein assays (SDS-PAGE gels and Western Analysis).

3. RESULTS

The results of the gene analysis component of this study are summarised in Table 2 and Table 3. Table 2 indicates that the data of the SYBR Green I analysis from the non-cancerous samples, positive control samples and cancerous samples do not have significant differences in melting peak temperatures. The control used in this experiment was lymphocyte derived DNA from individuals with no history of cancer.

Table 2. Melt peak analysis of NDUFB8.

NDUFB8	Melt Peak 1	Melt Peak 2
Control	81°C	79.8°C
Non-cancerous	79.48°C +/-1.08°C	78.67°C +/-0.98°C
Cancerous	79.2°C	79.07°C +/-0.63°C

Table 3 summarizes the data of the PTEN probe analysis. The Ct of Tubulin and PTEN in the non-cancerous sample are both within one cycle of each other, which is within the acceptable limits of housekeeper Ct variability. In the cancerous samples, it is seen that there is a three cycles difference when comparing the Ct of PTEN and Tubulin.

Table 3. Tubulin-α-8 (T) and PTEN (P) quantitative probe threshold cycle (Ct).

Sample	T Ct 1	P Ct 1	T Ct 2	P Ct 2	T Ct 3	P Ct 3	T Ct 4	P Ct 4
Non-cancerous	23.3±0.8	24.6±1.1	–	–	–	–	–	–
Cancer 1	29.5±0.5	26.8±0.9	33.3±0.5	30.2±0.5	22.4±0.1	19.8±2.6	27.4±0.5	26.1±0.4
Cancer 2	–	–	28.6±0.4	27.9±0.7	–	–	–	–
Cancer 3	–	–	30.5±0.5	28.9±0.8	–	–	–	–
Control	–	–	–	–	25.5±0.3	26.0±0.3	–	–

Table 4 is a summary of the PTEN protein concentrations in the HeLa cells after being treated by different concentrations of TZD. Using the lowest concentration as an internal standard, it is observed that as the TZD concentration increases, the concentration of PTEN increases as well.

Table 4. TDZ concentration and PTEN protein regulation.

TZD Concentration	0.125μM	0.25μM	0.5μM	1μM	2μM
PTEN Protein detected on PVDF in unit	9.8	16.1	15.6	25.6	21.6
Actin Protein detected on PVDF in unit	484.2	456.0	414.0	504.9	415.1
PTEN/Actin	2.02	3.51	3.76	5.07	5.21
(PTEN/Actin)/2.02	1	1.74	1.19	2.51	2.58

Figure 1 illustrates the up-regulation trend of PTEN protein as the concentration of TZD increases.

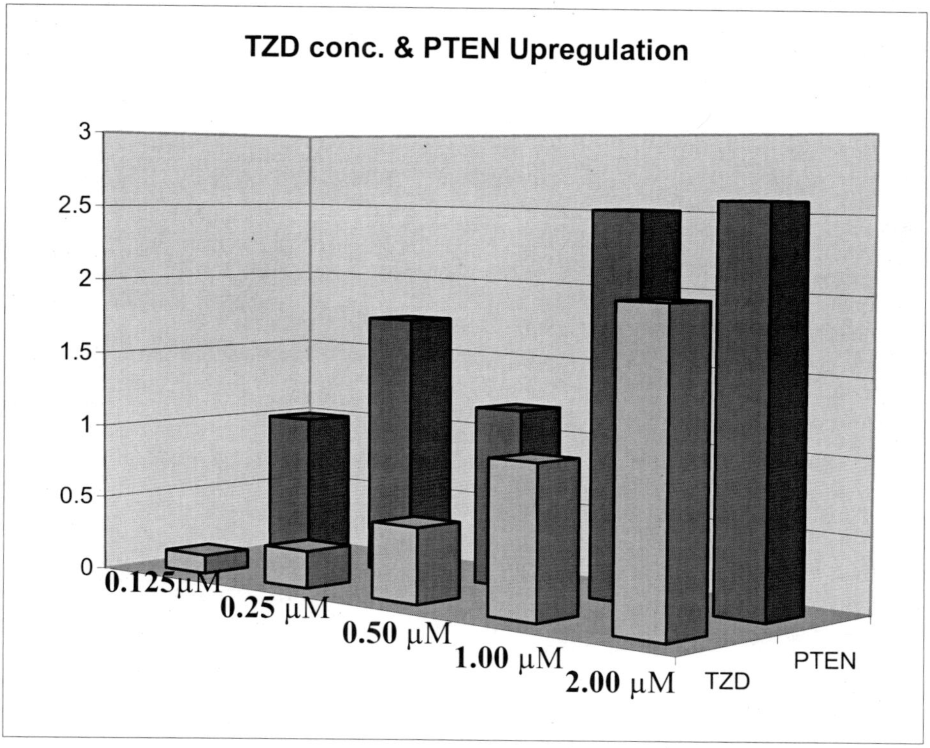

Figure 1. up-regulation trend of PTEN protein as the concentration of TZD increases.

4. DISCUSSION

This project aimed to investigate putative genomic alterations in cervical cancerous samples. Part of this project also aimed to investigate the involvement of a potential therapeutic anti-cancer drug using a cervical cancer cell line.

Several studies in our lab have indicated that nuclear-encoded genes of the ETC may play a role in skin carcinogenesis, which is a squamous epithelial cancer, like cervical cancer. This study analysed two aspects of mitochondrial regulation: a specific gene from the ETC (NDUFB8) and another gene (PTEN) localised to the same chromosomal region that plays an upstream role in the apoptotic pathway [2,3]. SYBR Green I Real-Time PCR was used to analyse the NDUFB8 gene. It is observed that the melting temperature from the non-cancerous samples, positive control samples and cancerous samples all have similar melting temperature. This implies that there are no major amplicon size differences between the non-cancerous and cancerous samples. The major limitation of the SYBR Green I experiment is the non-specific binding of SYBR Green I. SYBR Green I does not bind to a particular location on the DNA template, rather it binds to the major and minor grooves of the DNA molecule, respectively. The patent holder has not released the nature of this binding, thus the specificity of this experiment cannot be discussed. Furthermore, SYBR Green I reactions depend on the Magnesium concentration, which is pre-disposed to pipetting errors. This characteristic hampers between-run comparisons and sometimes within-run comparisons.

Real-Time PCR was performed on the PTEN and Tubulin α 8 genes using Dual-Labelled probe. In two of the three runs that were performed for this experiment, a minimum of two cycles difference was noted between the normal cervical tissue and the cervical cancerous tissue when a Ct comparison between the PTEN probe and the Tubulin α 8 (housekeeping gene) probe was performed. The third run listed in the result section has two samples existing at the border of being aberrant in terms of significant Ct differences. This run was performed under a different condition in that the pipetting was carried out on ice, due to the fact that some abnormalities existed in Ct differences (4 cycles difference) in one of the previous runs. It is known that keeping DNA and primers on ice can induce some lag time in the exponential phase of the Real-Time PCR reading. However, as the Tubulin α 8 housekeeping gene went through the Ct at the same cycle as in previous runs, the factor of performing pipetting on ice was discounted as being of major influence on the reaction conditions.

Due to the nature of how the Dual-Labelled probe binds to the template and the method of how the Rotor-Gene machine detects the increase in fluorescence, the two cycles difference noted in this project suggests a putative gain at this chromosome region. A chromosomal gain is indicative of a copy-number aberration, which is of the type detected in the Ashton et al.study of 2001. These types of aberrations can have wide-scale implications in regulating cellular functions. The major limitation of using Dual-Labelled probe Real-Time PCR is the requirement of an adequate amount of viable DNA recovered from the paraffin-embedded tissues. This can limit the sample number for the experimental analysis, and thus the scope of the conclusions.

In this section of the project, a putative up-regulation of the tumour suppressor PTEN protein was investigated. It is theorised that the ratio between the PTEN protein and β Actin protein (internal standard) was increased. This suggests a potential up-regulation of the PTEN protein after treatment of TZD.

The potential PTEN protein up-regulation after treatment of TZD has been studied in pancreatic cancer cell lines. However, the effect of TZD on PTEN regulation has not been studied in cervical cancer cell lines. Since it is known that TZD has the potential to up-regulate the PTEN protein, this study investigated the potential of whether a dose-dependent up-regulation of the PTEN protein in these cells existed. The results of this study indicated a PTEN up-regulation existed and that it may have a TZD does-dependent character. The major limitation of this experiment was the usage of a HeLa cell line, which is a transformed cell line that is known to have large-scale chromosomal abnormalities that can impact many cellular regulations and functions including the regulation of the expression of the PTEN protein. Also, this experiment has only been conducted once and thus identical studies are needed to make it statistically meaningful.

5. REFERENCES

1. Waggoner SE. Cervical Cancer. *Lancet* 361 (2003), pp. 2217-2225.
2. Chu EC and Tarnawski AS. PTEN regulatory functions in tumour suppression and cell biology. *Med. Sci. Monit.* **10** (2004), pp. RA 235-241.
3. Cheung TH, Lo KW, Yim SF, Chan LK, Heung MS, Chan CS, Cheung AY, Chung TK and Wong YF. Epigenetic and genetic alternation of PTEN in cervical neoplasm. *Gynecol Oncol.* **93** (2004), pp. 621-627.
4. Zeviani M and Di Donato S. Mitochondrial disorders. *Guarantors of Brain.* **127** (2004), pp. 2153-2172.
5. Lintell NA, Maguire DJ, McCabe M and Griffiths LR. Focussing on genomic and phenomic correlations in the respiration of non-melanomic skin cancers. *Adv. Exp. Med. Biol.* **566** (2004), pp. 375-380.
6. Ashton KJ, Weinstein SR, Maguire DJ, and Griffiths LR. Molecular cytogenetic analysis of basal cell carcinoma DNA using comparative genomic hybridisation. *Jf Invest. Dermatol.* **117**, No. 3 (2001), pp.683-686.
7. Grommes C, Landreth GE, and Heneka T. Antineoplastic effects of peroxisome proliferators activated receptor γ agonists. *Lancet Oncol.* **5** (2004), pp. 419-429.
8. Hong G, Davis B, Khatoon N, Baker SF and Brown J. PPAR gamma-dependent anti-inflammatory action of rosiglitazone in human monocytes: suppression of TNF alpha secretion is not mediated by PTEN regulation. *Biochem Biophys Res Commun.* **303(3)** (2003), pp. 782-787.

PREDICTION OF SURGICAL SITE INFECTIONS AFTER MAJOR SURGERY USING VISIBLE AND NEAR-INFRARED SPECTROSCOPY

Charlotte L. Ives*, D.K. Harrison*, and G.S. Stansby†

Abstract: Final results of an investigation into whether oxygen saturation of tissues (StO_2, measured by spectrophotometry) could predict surgical site infections (SSI) after major abdominal surgery are presented.

StO_2 was measured on the arm and wound site pre-operatively and then at 12, 24 and 48 hours post-operatively. A Whitland Research RM200 was employed as the visible lightguide spectrophotometer. StO_2 measurements using this machine were designated SSO_2 (skin SO_2). A Hutchinson Inspectra® Model 325 was used for the near infrared spectroscopy (NIS) measurements. StO_2 measurements using this machine were designated MSO_2 (muscle SO_2).

Of 59 patients (38 males, 21 females), 42 healed uneventfully and 17 developed SSI. The overall infection rate was 28.8%. No significant differences were seen in wound SSO_2 between outcome groups at any stage.

At 12 hours there was a significant difference between the two groups with respect to mean wound MSO_2 (A= 58.3+/-21.6%, B=42.2+/-16.6%, p=0.005, 95% confidence interval = 5.26, 26.98).

A receiver operating characteristic curve showed that when a wound MSO_2 of 53% was chosen as the threshold to classify potential infection a sensitivity of 71% and a specificity 73% (chi-squared test, p=0.002) was achieved.

The use of the near-infrared spectrophotometry as a tool to predict wound infections should be further evaluated and advocated.

* Medical Physics Department, University Hospital of North Durham, North Road, Durham, UK DH1 5TW

† Department of Surgery, Freeman Hospital, Newcastle upon Tyne, UK

1. INTRODUCTION

Wound infections (surgical site infections, SSI) are common, occurring in up to 30% of operations for abdominal surgery[1] and are estimated to cost the UK National Health Service up to £65 million per year[2]. Oxygen is the most important substrate for wound healing [3]; however, wounds by their very nature have a poor blood supply. Poor oxygenation leads to impaired bacterial killing by neutrophils as their bactericidal activity relies on the formation of oxygen radicals[4]. Oxygen is also needed for the rapid accumulation of collagen [5,6]. The mere act of increasing oxygen, either at wound site or systemically, has a beneficial effect on wound infection rates [7, 8, 9].

Increased levels of oxygen tension (pO_2) at the wound site have been shown to correlate with better healing and less infection, either measured subcutaneously[10] or transcutaneously [11, 12, 13]. As these methods of measuring pO_2 are invasive an alternative method of evaluating tissue oxygenation in patients undergoing surgery needs to be sought. This can be achieved using spectroscopy to measure tissue oxygen saturation (StO_2). Spectrophotometers shine light of known wavelengths onto the tissue under investigation, and by measuring the absorption spectra can calculate StO_2.

Peripheral oxygen saturation has been found to have some success in predicting healing after lower limb amputation[14]. Furthermore in a study to investigate the possible cause of the high infection rate in groin wounds following vascular bypass surgery Raza and colleagues[15] used the Erlangen microlightguide spectrophotometer[16] (EMPHO, BGT-Medizintechnik, Überlingen) to measure skin oxygen saturation (SSO_2) in the groin skin medially and laterally to the incision sites in patients undergoing femoro-popliteal or femoro-distal bypass operations prior to and at two and seven days post-operatively. The equivalent contralateral sites were used as controls. The results showed a significant difference ($p<0.01$) between the medial and lateral SSO_2 values post-operatively. On this basis, it was postulated that a disruption of blood supply may be responsible for the high incidence of infection in such surgical wounds.

It was proposed to use both lightguide and near-infrared spectroscopy in this study on the premise that StO_2 measured at different depths may provide an insight into the origins and depth of wound infections.

2. AIM

The main aim of this study was to evaluate whether StO_2 as measured by spectrometry could be used to predict wound infections. Last year we reported the results of a preliminary trial with the same aim, we now present our final results.

3. MATERIALS

A Whitland Research RM200 was employed as the visible lightguide spectrophotometer (LGS). StO_2 measurements using this machine were designated SSO_2 (skin SO_2).

For a near-infrared spectrophotometer (NIS) an Inspectra® tissue spectrophotometer Model 325 (kindly loaned by Hutchinson Technology Inc, Hutchinson, USA) was used. StO_2 measurements using this machine were designated MSO_2 (muscle SO_2).

Both of these machines have been previously described and used without adverse effects[17,18].

4. METHODS

After securing ethical approval from the Local Ethics Committee, patients undergoing major abdominal surgery were invited to participate. Only patients due to undergo major abdominal surgery were selected as they are at the highest risk of developing SSI. After gaining informed consent StO_2 was measured on the upper arm (10cm below shoulder tip over the bulk of biceps) and 2cm either side of the surgical site pre-operatively and then at 12, 24 and 48 hours post-operatively. Height and weight were also recorded to calculate body mass index (BMI).

Patients were assessed for signs of infection at the surgical incision at seven and 30 days post-operatively. This was performed by a healthcare professional blinded to the StO_2 readings using the Control of Diseases Centre guidelines for diagnosis of wound infection[19] Two groups were then classified retrospectively: those who healed uneventfully (Group A- "Non-infected") and those who developed SSI (Group B- "Infected").

5. RESULTS

All results with a Normal distribution will be expressed as mean ± standard deviation. Of 59 patients (38 males, 21 females), 42 healed uneventfully and 17 developed SSI. The overall infection rate was 28.8%. Mean age was 62.17 ± 16.31 years. There were no significant differences between the two outcome groups in terms of age, BMI or gender.

5.1. Arm StO_2

SSO_2 readings at each time interval were found to have a Normal distribution (Ryan-Joiner test for Normality), however it was found that at 24 hours post-operative MSO_2 readings were non-parametrically distributed and so all MSO_2 readings were analysed as being non-parametrically distributed.

As found in the preliminary trial[18] mean arm StO_2, as measured by either spectrophotometric technique, was not significantly different between the outcome groups at any stage (see table 1). MSO_2 results have been presented in a paper submitted elsewhere[20] and did not show any significant changes between groups or with time. For example pre-operatively: Group A, n=41, mean=59.06 ± 18.95%, group B, n=16, mean=57.50 ± 14.10, p=0.6. Males had significantly higher arm MSO_2 readings than females and this difference did not disappear when the readings were adjusted for BMI. Readings of arm MSO_2 at each time interval were negatively correlated with BMI (pre-operatively Pearson Correlation (PC) = -0.45, $p<0.001$). SSO_2 had a positive correlation with BMI (PC=0.41, p=0.002) but only for pre-operative readings.

Arm SSO_2 and MSO_2 were not correlated at any stage, muscle readings being significantly higher. For example pre-operatively: Arm SSO_2 compared to MSO_2: PC=0.04, p=0.76; Mann Whitney U-test, SSO_2 n=58, median=44.28%, MSO_2 n=57, median=63.67%, estimate for difference ($SSO_2 - MSO_2$) = -14.67%, $p<0.001$.

Table 1. Arm skin oxygen saturation readings shown overall and by outcome. Values are mean ± standard deviation. *indicates t-test

Timing	Overall	Non-infected	Infected	P value
Pre-operative (% saturation)	45.71 ±15.32 (n=58)	46.17 ± 16.00 (n=41)	44.61 ±13.92 (n=17)	0.71*
12 hours post-operative (% saturation)	48.20 ± 18.10 (n=53)	47.26 ±18.86 (n=36)	50.19 ±16.74 (n=17)	0.57*
24 hours post-operative (% saturation)	49.86 ± 15.34 (n=52)	48.76 ±15.96 (n=36)	52.34 ±14.00 (n=16)	0.42*
48 hours post-operative (% saturation)	45.60 ± 13.40 (n=51)	44.89 ±14.00 (n=36)	47.27 ±12.13 (n=15)	0.55*

5.2. Wound StO_2

All readings of wound StO_2 were found to be Normally distributed.

No significant differences were seen in wound SSO_2 between outcome groups at any stage (see table 2). However wound SSO_2 rose significantly in all patients after the operation at 12 and 24 hours post-operatively; when this was analysed by outcome groups it was only significant at 24 hours in patients who developed SSI.

Table 2. Mean wound SSO_2 readings over time, shown overall and by outcome. Values are mean ± standard deviation. *indicates t-test

Timing	Overall	Non-infected	Infected	P value
Pre-operative (% saturation)	38.02 ±11.71 (n=58)	38.20 ± 12.25 (n=41)	37.58 ± 10.66 (n=17)	0.85*
12 hours post-operative (% saturation)	42.35 ± 12.59 (n=52)	41.53 ± 13.79 (n=35)	44.06 ± 9.82 (n=17)	0.45*
24 hours post-operative (% saturation)	42.84 ± 13.05 (n=50)	41.34 ± 13.49 (n=34)	46.00 ± 11.84 (n=16)	0.23*
48 hours post-operative (% saturation)	40.72 ± 14.61 (n=51)	39.64 ± 14.38 (n=36)	43.29 ± 15.33 (n=15)	0.44*

MSO_2 results have been presented in a paper submitted elsewhere[20] but are briefly discussed below:

Results found in the preliminary trial[18] have been upheld; at 12 hours post-operatively there is still a significant difference between the two outcome groups with respect to mean wound MSO_2 (Group A: n=38, 55.81 ± 21.97%, Group B: n=17, 43.40 ± 18.07%, p=0.03, 95% confidence interval = 0.96 to 23.85%). Chi-squared tables for differing values of wound MSO_2 at 12 hours post-operatively showed that a value of 53% chosen as the threshold to classify potential infection was the most effective (chi-squared test, p=0.002). If MSO_2 values at 12 hours post-operatively were to be used as a predictive test this chosen level would correlate to a sensitivity of 70.6% and a specificity of 76.2%.

Abdominal SSO_2 and MSO_2 were not correlated at any stage, muscle readings being significantly higher. For example pre-operative values: PC=0.19, p=0.18; Two-sample t-test, SSO_2 n=58, mean=38.0±11.7%, MSO_2 n=55, mean=42.9±21.2%, 95% CI for difference ($SSO_2 - MSO_2$) = -11.38 to +1.56%, p=0.14.

No correlations were seen between abdominal SSO_2 and BMI. Abdominal MSO_2 was significantly negatively correlated with BMI at each time interval. For example, pre-operatively PC= -0.64, p<0.001.

5.3. Skin oxygen saturation compared between sites

Unlike the preliminary study[18] a strong correlation between SSO_2 at sites was found pre-operatively: PC=0.69, p< 0.001. This correlation was also present at 24 and 48 hours post-operatively (PC=0.59, p<0.001 and PC=0.43, p=0.002, respectively) but not at 12 hours post-operatively (PC=0.26, p=0.07).

5.4. Muscle oxygen saturation compared between sites

As in the preliminary study[18] MSO_2 is related between sites. Pre-operative readings were compared first and were found to have a strong correlation – PC=0.61, p<0.001.These correlations continued throughout the post-operative period.

5.5. Degrees of tissue hypoxia

Calculating the degree of tissue hypoxia is a method that has been used to predict stump healing following lower limb amputation for critical ischaemia[14]. It is defined as the percentage of values below a predefined level and is a useful measurement because SSO_2 can be very variable. The degree of tissue hypoxia is then plotted against the mean SSO_2, the theory being those patients with high degrees of tissue hypoxia and a low mean tissue oxygenation will be less likely to heal.

At each time interval in this study both mean arm and abdominal SSO_2 were analysed against varying degrees of tissue hypoxia (set at values below 10, 15, 20, 25, 30, 35 and 40% SSO_2). To display all 56 graphs of degrees of tissue hypoxia against mean SSO_2 is beyond the scope of this paper, instead an example is produced in Figure 1. This shows the 'hypoxia level' set at 35% for abdominal SSO_2 at 12 hours post-operatively. Each point on the graph represents one patient – the number of abdominal SSO_2 readings below 35% plotted against the mean abdominal SSO_2 at that time. Patients who developed a wound infection were marked as a white circle; those patients who healed uneventfully were marked as a black triangle.

No time interval or level of hypoxia found a perfect fit for the data. A perfect fit would be one where all of the infected outcome group data were clustered together in the lower right hand corner. In figure 1 there are some patients with very low degrees of tissue hypoxia who develop SSI, and some patients with very high levels of hypoxia who heal uneventfully.

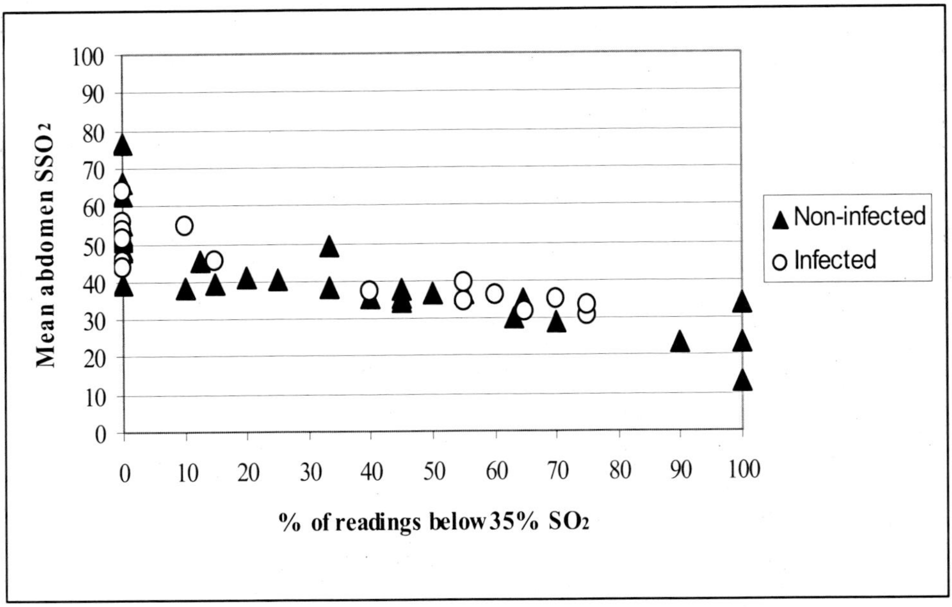

Figure 1. An example of degrees of tissue hypoxia test at 12 hours post-operatively. Mean abdominal SSO_2 is plotted against tissue hypoxia set at a level of below 35%.

Values for degrees of tissue hypoxia were non-Normally distributed and so were compared using a Mann-Whitney U test. For both arm and abdominal SSO_2 comparisons were made between outcome groups at each time interval, and within outcome groups from pre-operative to post-operative levels. All calculations showed there were no statistically significant differences between outcome groups at each time interval or within outcome groups at different time intervals.

6. DISCUSSION

Unlike the study by Hopf and colleagues this study found that arm tissue oxygenation was not useful as a predictive tool for SSI. This is not surprising for SSO_2 as tissue oxygenation is being measured at a different level, i.e. the skin rather than subcutaneous tissue. Furthermore it has been shown in previous studies that skin has a blood supply which appears to be controlled locally[17, 18]. Interestingly correlations were seen in this study between the two areas, this could be a chance finding or could represent there is maximal vasodilation due to high ambient temperatures found on most hospital wards. No difference seen between outcome groups in arm MSO_2 could be explained by the fact that no wound was created in the arm and so we were not measuring the dynamic changes that occur when wounds heal.

Regarding abdominal StO_2, SSO_2 was not significantly different between the two outcome groups; this could again be explained because of the heterogeneity of skin blood supply. The 'degrees of hypoxia' analyses were not useful because the tissue under study is not critically ischaemic, unlike the amputation levels assessment. Post-operatively the responses to surgery will falsely elevate the readings, especially if infection is present; abdominal SSO_2 rose significantly at 24 hours post-operatively in infected patients.

The most significant finding in this study was the difference in abdominal MSO_2 values between outcome groups at 12 hours post-operatively. This also represented a significant mean rise for non-infected patients from the mean baseline value at the same time interval. This again shows the body's response to surgery; there is an increase in oxygen delivery to meet the metabolic demands for healing. It may be that those who develop infection have an inappropriate response and shunting of oxygen to the wrong areas, or the increase in perfusion is inadequate. However no trend in post-operative changes or percentage change was attributable to either outcome group.

At either site there was no correlation between SSO_2 and MSO_2, this shows the two spectrophotometric techniques are measuring different tissue depths. MSO_2 is negatively influenced by BMI – a measurement of adiposity. This probably means that the NIRS is actually measuring SO_2 of the subcutaneous fat layer. This is not a problem for predicting wound infections as the poor blood supply of fat is the determining factor for developing SSI – most wound infections originate from this layer.

7. CONCLUSION

NIRS appears to measure StO_2 of subcutaneous adipose tissue. Our label for the data obtained by the NIRS should be changed from MSO_2 to a more appropriate label, for example ASO_2 (adipose).

Tissue oxygenation, as measured by near-infrared spectrophotometry, could potentially be a useful tool to predict wound infections. The use of such a tool should be further evaluated to assess whether tissue oxygenation can be improved at 12 hours post-operatively, and whether this affects outcome.

8. REFERENCES

1. L.A .Israelsson. The surgeon as a risk factor for complications of midline incisions. *Eur J Surg* 1998; **164**(5): 353-9.
2. R. Plowman, N. Graves, M. Griffin et al. The socio-economic burden of hospital acquired infection. *Public Health Laboratory Service* 2000
3. T.K. Hunt, B. Zederfeldt, T.K. Goldstick. Oxygen and healing. *Am J Surg* 1969; **118**: 521-5
4. D. Allen, J. Maguire, M. Mahdavian et al. Wound hypoxia and acidosis limit neutrophils bacterial killing mechanisms. *Arch Surg* 1997; **132**(9): 991-996
5. T.K. Hunt, M.P. Pai. The effect of varying ambient oxygen tensions on wound metabolism and collagen synthesis. *Surg Gynecol Obstet* 1972; **135**: 561-7
6. K. Jonsson, J.A. Jensen, W.H. Goodson III et al. Tissue oxygenation, anemia, and perfusion in relation to wound healing in surgical patients. *Ann Surg* 1991; **214**(5): 605-13
7. J.H.Niinikoski. Clinical hyperbaric oxygen therapy, wound perfusion, and transcutaneous oximetry. *World J Surg* 2004; **28**(3): 307-11
8. R. Greif, O. Akça, E.P. Horn et al. Supplemental perioperative oxygen to reduce the incidence of surgical-wound infection. *N Engl J Med* 2000; **342**(3):161-7
9. D.R. Knighton, B. Halliday, T.K. Hunt. Oxygen as an antibiotic. The effect of inspired oxygen on infection. *Arch Surg* 1984; **119**(2):199-204
10. H.W. Hopf, T.K. Hunt, J.M. West et al. Wound tissue oxygen tension predicts the risk of wound infection in surgical patients. *Arch Surg* 1997; **132**: 991-6
11. J. McMahon, M. Grigg. Predicting healing of lower-limb ulcers. *Aust NZ J Surg* 1995; **65**(3):173-76
12. F.T. Padberg, T.L. Back, P.N. Thompson et al. Transcutaneous oxygen (TcPO2) estimates probability of healing in the ischemic extremity. *J Surg Res* 1996; **60**: 365-369
13. M. Kalani, K. Brismar, B. Fagrell et al. Transcutaneous oxygen tension and toe blood pressure as predictors for outcome of diabetic foot ulcers. *Diabetes Care* 1999; **22**(1):147-51
14. D.K. Harrison, P.T. McCollum, D.J. Newton, et al. Amputation level assessment using lightguide spectrophotometry. *Prosthet Orthot Int* 1995;**19:**139-47

15. Z. Raza, D.J. Newton, D.K. Harrison et al. Disruption of skin perfusion following longitudinal groin incision for infrainguinal bypass surgery. *Eur J Vasc Endovasc Surg* 1999; **17**: 5-8
16. K.H. Frank, M. Kessler, K. Appelbaum. The Erlangen micro-lightguide spectrophotometer EMPHO I. *Phys Med Biol* 1989; **34**: 1883-1900
17. C.L. Ives, D.K. Harrison, G. Stansby. Relationships between muscle SO_2, skin SO_2 and physiological variables. *Adv Exp Med Biol* 2005, in press
18. C.L. Ives, D.K. Harrison, G. Stansby. Prediction of surgical site infections using spectrophotometry: preliminary results. *Adv Exp Med Biol* 2005, in press
19. T.C .Horan, R.P. Gaynes, W.J. Martone, et al. CDC definitions of nosocomial surgical site infections, 1992: a modification of CDC definitions of surgical wound infections. *Am J Infect Control* 1992; **20:** 271-274
20. C.L. Ives, D.K. Harrison, G. Stansby. Prediction of surgical site infections after major surgery using spectrophotometry, submitted to *BJS* 2006

APPLICATION OF NOVEL METAL NANOPARTICLES AS OPTICAL/THERMAL AGENTS IN OPTICAL MAMMOGRAPHY AND HYPERTHERMIC TREATMENT FOR BREAST CANCER

Hanzhu Jin and Kyung A. Kang*

Abstract: Low heat (42~45°C) hyperthermia is an effective cancer therapy with few side effects. Well engineered nanoparticles can effectively guide heat to the tumor without damaging the normal tissue. When nano-sized particles are injected to an organ with a tumor, they tend to accumulate in the tumor due to the unorganized nature of its vasculature. Magnetic nanoparticles, such as Fe_3O_4, are heated by a well selected alternating electromagnetic frequency. At a frequency of 450 KHz or lower, Fe_3O_4 nanoparticles at a size of 10~30 nm were heated effectively, without heating tissue main components. Gold nanoparticles are known as strong near infrared (NIR) absorbers. Nanogold particles at a diameter of 150 nm were added at 0.01wt% to the tumor model, placed at depths of 1~2.5 cm, in an optically equivalent experimental breast model. Then the surface of the breast model was scanned with NIR light at 788 nm. The particles in the tumor model increased the optical contrast of the tumor by 1~3.5dB. Considering that some of the FDA approved MRI contrast agents are made of Fe_3O_4, gold-coated Fe_3O_4 particles have a potential to be used as safe optical and thermal markers, allowing seamless breast cancer detection and cancer-specific hyperthermic treatment.

1. INTRODUCTION

Cancer specific, low heat hyperthermia is a useful cancer treatment method with minimal damage to the normal tissue.[1] Magnetic nanoparticles have been previously considered to deliver the heat to destroy the tumor under an alternating electromagnetic (AEM) field.[2,3] One of the most frequently used magnetic materials for hyperthermia is iron oxide nanoparticle (Fe_2O_3 or Fe_3O_4).[4,5] Considering that the Food and Drug

* Hanzhu Jin and Kyung A. Kang, Department of Chemical Engineering, University of Louisville, Louisville, KY 40292.

Administration (FDA) has approved iron oxide based MRI contrast agents for human use (Feridex I.V.® or Endorem®; Advanced Magnetics, Cambridge, MA), the toxicity of iron oxide nanoparticles may be minimal.

Near infrared (NIR) imaging has been very effective in breast cancer detection, sometimes, even for some tumors that are not detected by X-ray.[6] When a tumor is, however, very small and deeply seated, the optical contrast by the tumor can be weak, and safe optical contrast agents can be very effective. Gold nanoparticles are effective contrast agents because they are chemically inert and absorb NIR well.[7] The optical resonance wavelength of gold particles can be easily tuned to a broad range, especially to the NIR range, by controlling their sizes and shapes.[8, 9, 10] Hong and Kang have also demonstrated that gold nanoparticles at an appropriate size enhance the fluorescence signal when a fluorophore is linked to the nanoparticle at a proper distance.[11] Nano-sized particles are smaller than capillaries (the diameter of usual mammary capillary is approximately 7 μm)[12] and therefore, are easily circulated within the blood vessels. When these nanoparticles are injected to an organ with a tumor, they tend to accumulate in the tumor due to the unorganized vasculature of tumors. Tumor specific antibodies can also be linked to the particles and can better guide them to the tumor.

Therefore, gold-coated magnetic nanoparticles can be used as dual purposed, optical/thermal agents: the gold layer for high NIR absorption or scattering, and the magnetic core for guiding the heat to the tumor, both non-invasively. In this study, the heating performance of Fe_3O_4 nanoparticles as possible thermal agents for breast cancer hyperthermic treatment was investigated. Also, a preliminary study using gold nanoparticles as NIR contrast agents for breast cancer detection has been conducted.

2. MATERIALS AND METHODS

2.1. Hyperthermia Study

Induction heaters (MKII-5; Taylor-Winfield Induction Company; Brookfield, OH) at frequencies of 0.4 ~ 10 MHz and a power of 5 KW were used. The induction heater was made of a 0.5 cm thick copper tube, a coil of three turns, at a dimension of 2 cm (diameter) x 3 cm (height). γ-Fe_2O_3 and Fe_3O_4 at various sizes (Alfa Aesar; Ward Hill, MA) were uniformly dispersed in 4 ml of agar gel (2.15wt%: weight percentage) at various concentrations. The mixture containing the particles was placed into a glass test tube of 1.6 (diameter) x 10 (length) cm, and solidified at room temperature. The glass tube was then placed inside the induction heater coil [Figure 1(a)] and the AEM field was applied for a predetermined period. The sample temperature was measured by Traceable® digital thermometer (accuracy: ± 0.05°C; All QA product; Belmont, NC) before and after applying the AEM field.

For the preliminary test of non-invasive heating, an empty Vitamin E capsule shell (2.5 x1.0x1.0 cm; National Vitamin; Las Vegas, NV) was filled with 1wt% Fe_3O_4 and 2.15wt% agar and solidified. It was then placed at 1 cm depth in a cylindrically shaped tissue model [10.0 (diameter) x 7.0 (length) cm] made of agar and water. After the tissue model was solidified, a ring-shaped [inner diameter of 10 cm with 3 turns, Figure 1(b-1)] or pancake-shaped [3 or 10 cm diameter with 4 turns, Figure 1(b-2)]] induction heater was placed on the surface of the tissue model and the field was applied to observe the heating of the tissue and the tumor models.

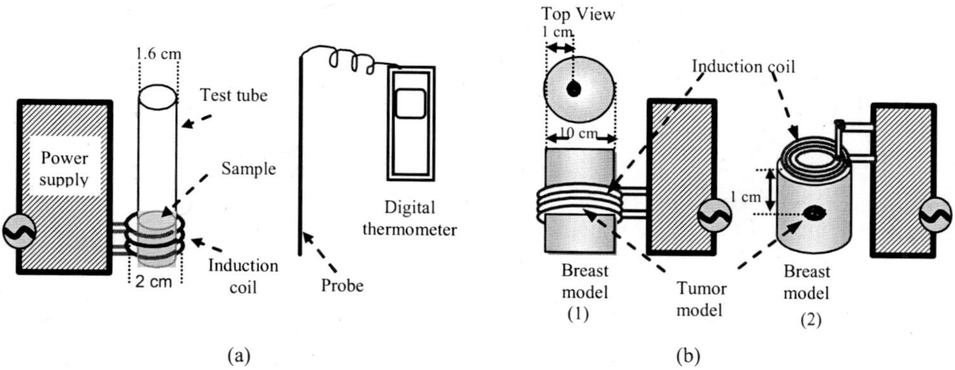

Figure 1. Schematic diagram of the experimental set up for (a) heating dispersed magnetic nanoparticles in agar and (b) non-invasive heating of magnetic nanoparticle containing tumor model by (1) the ring-shaped and (2) pancake-shaped induction coils in the experimental tissue model.

2.2. Near Infrared Spectroscopy Study

Optically-equivalent breast tissue like models of 24.0 x 14.0 x 5.0 cm were constructed following the procedure described by Jin and Kang.[13] Briefly, the absorption and scattering properties of the breast model were adjusted by using India ink and skimmed milk (absorption coefficient: ~ 0.03 cm^{-1}; scattering coefficient: ~ 6 cm^{-1}). Gold nanoparticles (Ted Pella, Inc.; Redding, CA) at a size of 150 nm were added (0.01wt%) to the same ingredient used for the breast model and the mixture was injected into an empty Vitamin E capsule shell (1.5 x 1.0 x 1.0 cm). Once the capsule was solidified, it was placed in the breast model at predetermined depths (1 and 2.5 cm).

After solidification of the breast model, the surface of the model was scanned by the optical fibers of the NIR - time resolved spectroscopy (TRS) instrument for absorption both in reflectance and transmittance. The instrumentation was described by Honar and Kang.[14] The measurements were performed at an interval of 1 cm on an area of 5.0 x 5.0 cm around the absorber. Obtained TRS spectra were converted to the frequency domain using a computer code for Fourier transformation.[15]

3. RESULTS AND DISCUSSION

3.1. Heating Performances of Magnetic Nanoparticles in the AEM Field

3.1.1 Selection of Safe AEM Frequencies

To select the AEM frequency range that can heat nanoparticles accumulated in/around the tumor, without heating normal tissue, bovine hemoglobin solutions at a concentration of 0.14 g/ml-water (normal blood), 0.9% NaCl solution, and ground beef were placed in the AEM field at various frequencies. 4 ml of each sample was placed inside the induction heater coil [Figure 1(a)], at frequencies of 0.45, 5.4, and 9.2 MHz, at 5 KW for 2 minutes (Table 1).

Table 1. The heating of tissue components by an AEM field at a frequency range of 0.4~10 MHz, with a 5 KW power, for 2 minutes.

Samples	Concentration	ΔT (°C) 0.45 MHz	5.4 MHz	9.2 MHz
Distilled H_2O		2.7	3.0	2.9
Hemoglobin	0.14 g/ml	3.2	1.8	-*
NaCl Solution	0.9%	3.1	12.7	11.2
Ground beef		3.2	5.9	8.3

* Not tested.

The distilled water was heated minimally at all frequencies. Hemoglobin also did not become heated at the frequencies tested. 0.9% (physiological concentration) of NaCl solution was heated extensively at 5.4 and 9.2 MHz. The ground beef was also heated at 5.4 and 9.2 MHz, probably due to the salt content in the meat. The frequency of 0.45 MHz was, therefore, selected for further studies.

3.1.2 Heating Effect of Iron Oxide Nanoparticles in AEM

As a next step, the effect of nanoparticle size on heating performance was studied at 0.45 MHz and 5 KW. Figure 2 shows the temperature increase in the samples (in agar) containing the iron oxide nanoparticles at various sizes, after exposure in the AEM field for 2 minutes. The concentrations of iron oxide nanoparticles in the sample were ranged from 0.1 to 1wt%. For all particles the heating was linearly proportional to the particle concentration in the sample. For the samples containing Fe_3O_4 at 10-20 nm or 20-30 nm, the temperature increases were approximately at a rate of 35°C/1wt% of particles. With Fe_3O_4 particles at 40-60 nm, however, it was 9°C/1wt%, indicating that the heating capability of Fe_3O_4 nanoparticles depend on their particle sizes.

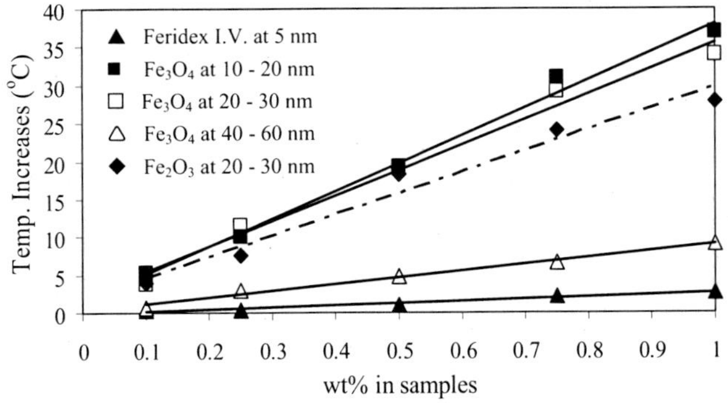

Figure 2. The effect of particle size on heating performance of iron oxide nanoparticles by AEM field. The concentrations of iron oxide in the samples were 0.1 ~ 1wt%. The samples were heated at a 0.45 MHz frequency and 5 KW power for 2 minutes.

For the sample containing 20-30 nm size of γ-Fe_2O_3, the rate of temperature increase was 30°C/1wt%, which is very close to the Fe_3O_4 particles at the range of 10 ~ 30 nm. For the Feridex I.V.®, which is a 5 nm Fe_3O_4 based MRI contrast agent, the rate of temperature increase was 3°C/1wt%. Although the heating rate is rather slow, considering it is FDA approved and only 2 minutes of heating increases 3°C, extending the AEM application time can achieve the desired temperature.

3.1.3 The effect of geometry of induction coil on heating

As a next step, the feasibility of non-invasive heating of tumor by the AEM field was studied. A tumor model, containing Fe_3O_4 particles, was placed at 1 or 2.5 cm depth in a cylindrical shaped tissue model made of agar and water as previously stated. The tumor model is an empty vitamin E capsule shell filled with 1wt% of 10~20 nm size of Fe_3O_4 in agar gel. A ring- or pancake-shaped induction coil was applied around/on the breast models [Figure 1(b)].

Table 2 shows the effect of the geometry of the induction coil on the tumor model containing iron oxide nanoparticles in the tissue model. With the ring-shaped induction coil at a diameter of 10 cm, after 10 minutes of heating, very little heating was observed for the tumor at 1 or 2.5 cm depth. With the pancake-shaped coil at a diameter of 10 cm, little temperature increase was observed after 10 minutes for the tumor at both 1 and 2.5 cm depth. For the pancake-shaped induction coil at a diameter of 3 cm, however, the tumor temperature at 1 cm depth increased by 17°C after only 2 minutes and even at 2.5 cm depth, the tumor temperature increased by 1°C. In all cases, no temperature increase was observed for the control. Therefore, the geometry of the induction coil (in other words, the AEM field strength) is a very important factor to be considered to optimize the heating of nanoparticles.

Table 2. Effect of the geometry of the induction coil on nanoparticle heating.

Induction coils	Diameter (cm)	Heating time (Min.)	ΔT (°C)	
			1 cm deep	2.5 deep
Ring Shape	10	10	2	0
Pancake Shape	10	10	0	0
	3	2	17	1

3.1.4 Optimization of geometries of induction coils

To study the AEM field penetration depth in breast tissue, computer simulations of the AEM field were performed using a VIZIMAG computer code by John Stuart Beeteson (Academic Press; London, UK). The field around a single and a dual pancake type coil(s) were simulated (Figure 3). The level of brightness shows the density of the magnetic flux. For the single pancake-shaped coil, the flux density seems to be distributed along the coils, with little diffusing to the neighboring areas [Figure 3(a)]. For the two pancake system, with the same current flow directions for both, the magnetic flux density between the coils are greatly enhanced [Figure 3(b)].

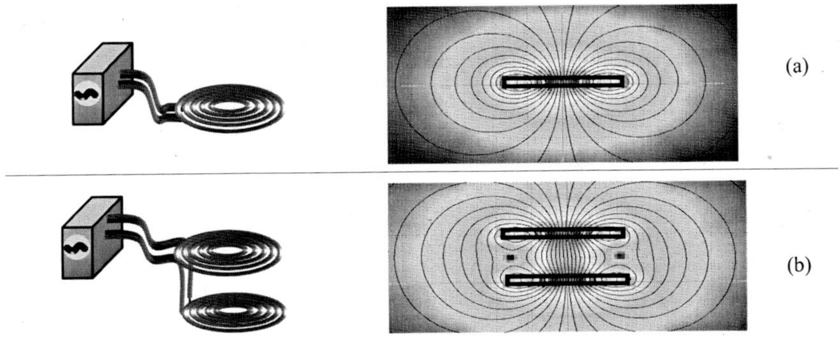

Figure 3. Schematic diagram of (a) single and (b) double pancake-shaped induction heater coils and the simulation results of the magnetic field distribution (front view).

3.2. Enhancement of absorption contrast by gold nanoparticles

From our study of light absorption by various sized nano-gold particles (5 ~ 250 nm), the particles at a size of 150 nm have shown the highest absorption at a wavelength of 788 nm. To test the optical contrast by the 150 nm sized gold particles, they were added to the breast model ingredients at a concentration of 0.01wt% and filled in the vitamin E capsule shell. The solidified tumor model was placed in the breast model at a depth between 1 ~ 2.5 cm. Figure 4 shows the magnitude ratio of light intensity (ΔMR) at a modulation frequency of 100 MHz, obtained from the area (5.0 x 5.0 cm) around the tumor location, when the measurement was taken in transmittance. For the 1 cm deep tumor model, the maximum decrease of MR was about 3.5 dB. Even at 2.5 cm deep, it showed the maximum contrast of 1 dB, demonstrating that the gold nanoparticles are effective contrast agents for NIR.

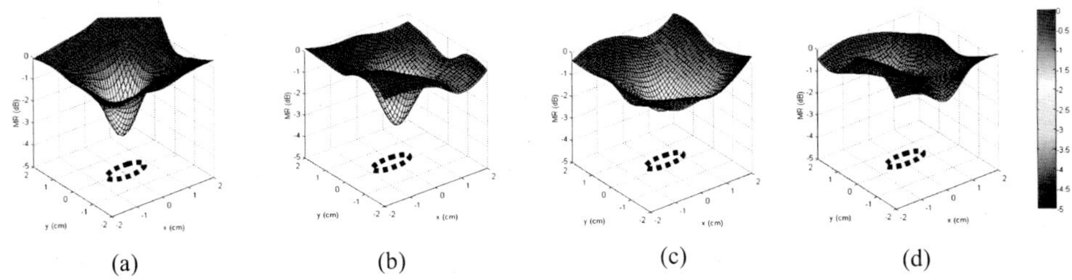

Figure 4. ΔMR in transmittance when a tumor model containing gold nanoparticles is located at (a) 1.0, (b) 1.5, (c) 2.0, and (d) 2.5 cm in the breast model at the modulation frequency of 100 MHz. The black dashed ellipses are the actual sizes and positions of the tumor models.

From the results of Fe_3O_4 particle heating and the optical contrast of gold particles, it is expected that gold coated Fe_3O_4 particles can be used as dual functional nano-entity, i.e., optical and thermal marker. Biocompatible surfactant layers, i.e. acrylic acid, may be coated on the gold coated Fe_3O_4 to make the particles hydrophilic.

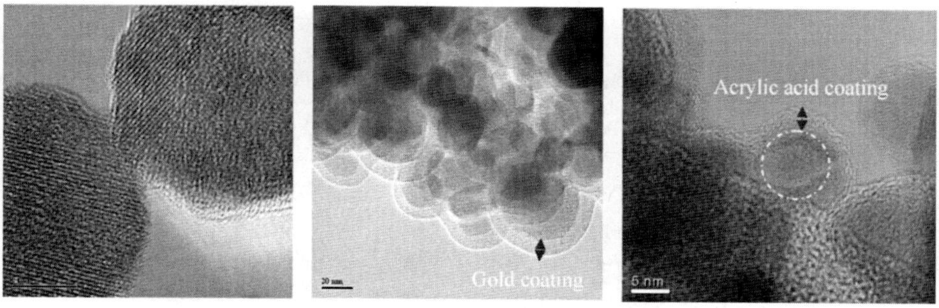

(a) (b) (c)

Figure 5. TEM images of (a) Fe_3O_4, (b) gold coated Fe_3O_4, and (c) acrylic acid coated Fe_3O_4. [The TEM images of (a) and (b) were taken by our group, and (c) is provided Dr. Shi in the University of Cincinnati, OH.].

Figure 5 shows the transmission electron microscopy (TEM) images of (a) Fe_3O_4, (b) gold coated Fe_3O_4, and (c) acrylic acid coated Fe_3O_4 nanoparticles. Fe_3O_4 was coated with gold using the wet solution method by Shi et al. at the University of Cincinnati, OH. The particles were also coated with acrylic acid using plasma coating by the same group, allowing particles to dissolve in water at a concentration of approximately 0.1wt%.

4. CONCLUSIONS

From the studies performed on nanoparticles for their heating by the AEM field and generation optical contrast, the following is concluded: Iron oxide nanoparticles were easily heated by the AEM field at a frequency of 0.45 MHz, whereas the main tissue components were not heated. Nanoparticles can be used as a heat guide for tumor hyperthermic treatment, and their heating capability depends on the size. The particle size range of 10-30 nm seems to get heated well. Our study results on heating a tumor model containing nanoparticles showed that non-invasive hyperthermic treatment may be feasible. The geometry of the coil was found to be a very important factor. Gold nanoparticles demonstrated that they enhance the absorption of NIR. Therefore, if Fe_3O_4 nanoparticles are coated with gold, they can be used as optical/thermal markers for seamless detection and treatment of breast cancer in a minimally invasive way.

The future studies include the gold coating on the Fe_3O_4 nanoparticles and testing of their optical/thermal properties and induction heater coil design for an optimally functional magnetic field generation for human breast.

5. AKNOWLEDGEMENT

The authors thank the U.S. Army Medical Research and Materiel Command (DAMD17-03-1-0572). The authors thank Dr. Shi's group at the University of Cincinnati, OH for supplying the gold or acrylic acid coated nanoparticles. The authors would like to express their appreciation to Topworx Company (Louisville, KY) for allowing them to use an induction heater.

6. REFERENCES

1. F. Kristian Storm, *Hyperthermia in Cancer Therapy*, (G. K. Hall Medical Publishers, Boston, MA, 1983).
2. A. Jordan, R. Scholz, P. Wust, H. Fahling, and R. Felix, , Magnetic fluid hyperthermia (MFH): cancer treatment with AC magnetic field induced excitation of biocompatible superparamagnetic nanoparticles *J. Magn. Magn. Mater.* 201 413–19 (1999).
3. P. Tartaj, M.P. Morales, S. Veintemillas-Verdaguer, T. Gonz´alez-Carreno, and C.J. Serna, The preparation of magnetic nanoparticles for applications in biomedicine, *J. Phys. D: Appl. Phys.* 36, R182–197 (2003).
4. D. Bahadur, and J. Giri, Biomaterials and magnetism, *Sadhana*, 28 (3 and 4), 639 – 656 (2003).
5. S. Mornet, S. Vasseur, F. Grasset, and E. Duguet, Magnetic nanoparticle design for medical diagnosis and therapy, *J. Mater. Chem.*, 14, 2161 – 2175 (2004).
6. M. A. Franceschini, K. T. Moesta, S. Fantini, G. Gaida, E. Gratton, H. Jess, W. W. Mantulin, M. Seeber, P. M. Schlag, and M. Kaschke, Frequency-domain techniques enhance optical mammography: Initial clinical results, *Proc. Natl. Acad. Sci. USA* 94, 6468-6473, (1997).
7. C.F. Bohren, and D.R. Huffman, *Absorption and Scattering of Light by Small Particles* (Wiley: New York, 1983).
8. S. J. Oldenburg, J. B. Jackson, S. L. Westcott, and N. J. Halas, Infrared extinction properties of gold nanoshells, *Applied Physics Letters*, 78 (19), 2897-2899, (1999).
9. C. H. Chou, C. D. Chen, and C. R. Wang, Highly Efficient, Wavelength-Tunable, Gold Nanoparticle Based photothermal Nanoconvertors, *J. Phys. Chem. B*, 109, 11135-11138, (2005).
10. J. Chen, F. Seaki, B.J. Wiley, H. Gang, M.J. Cobb, Z. Li, L. Au, H. Zhang, M.B. Kimmey, X. Li, and Y. Xia, Gold Nanocages: Bioconjugation and Their Potential Use as Optical Imaging Contrast Agents, *Nano Letters*, **5** (3), 473-477, (2005).
11. B. Hong and K.A. Kang, Biocompatible, nanogold-particle fluorescence enhancer for fluorophore mediated, optical immunosensor, *Biosensors and Bioelectronics*, 21(7), 1333-1338, (2006).
12. R. Eckert, D. Randall, G. Augustin, *Animal Physiology*, 3rd Edition, W. H. Freeman and Company, New York, P435-473, (1988).
13. H. Jin and K.A. Kang, Fluorescent mediated detection of Heterogeneity in a highly scattering media, *Adv. Exper. Med. Bio.*, 566, 167-172 (2005).
14. A.L. Honar and K.A. Kang, Effect of the source and detector configuration on the detectability of breast cancer, *Comp. Biochem.Physio - Part A: Molecular & Integrative Physiology,* 132(1), 9-15 (2002).
15. D. F. Bruley, Pulse reduction code written for process identification (personal communication), 1974.

POSSIBLE MECHANISMS OF IMPROVED RADIATION RESPONSE BY CYTOTOXIC RNASE, ONCONASE®, ON A549 HUMAN LUNG CANCER XENOGRAFTS OF NUDE MICE

Dae Hong Kim[1], Eun Ju Kim[1], Anna Kalota[2], Alan M. Gewirtz[2], Jerry Glickson[1], Kuslima Shogen[3], and Intae Lee[1,4] *

Abstract: The cytotoxic RNase, Onconase® (ONC), isolated from amphibian oocytes, was used to study its effect on the radiation response in A549 human NSCLC *in vitro* and *in vivo*. In cell culture studies, we found that ONC increased the radiation response by ONC-induced inhibition of O_2 consumption (QO_2). The occurrence of apoptosis was increased by ONC and was dependent on dosages and time exposure (measured by a Tunnel *in situ* cell death detection assay). Moreover, ONC inhibited sublethal damage repair (SLDR), confirmed by a split dose experiment. In animal studies, ONC significantly increased the radiation-induced tumor growth delay of A549 tumors *in vivo*. Using a non-invasive DCE-MRI technology, ONC-induced changes of perfusion were observed in A549 tumors. We concluded that the ONC-induced enhancement in tumor oxygenation was mainly due to the reduction in QO_2 rather than an increase in tumor blood flow. This investigation suggests important potential clinical uses of ONC for the treatment of NSCLC cancer patients.

1. INTRODUCTION

In the United States alone more than 170,000 people are affected by malignant lung cancer. This very heterogeneous type of malignancy is divided into two categories, small

* [1]Department of Radiology and [2]Department of Medicine (Division of Hematology/ Oncology), University of Pennsylvania, Philadelphia, PA 19104, USA: [3]Alfacell Corporation, Bloomfield, NJ 07003, USA.

[4]To whom correspondence and reprint requests should be addressed at Department of Radiology, University of Pennsylvania, B6 Blockley Hall, 423 Guardian Drive, Philadelphia, PA 19104-6069, USA. Telephone: 215-898-1805, Fax: 215-573-2113, Email: leeint@uphs.upenn.edu

cell lung carcinoma (SCLC) and non-small cell lung carcinoma (NSCLC). NSCLC is the most common type of lung cancer, comprising about 80% of all cases. Currently, Onconase (ONC: RNases from oocytes of *Rana pipiens*) is in randomized Phase III trials (ONC + doxorubicin vs. doxorubicin) for patients with malignant mesothelioma (1). The Phase I/II trial of ONC has been initiated in patients with NSCLC.

Since no publications have observed any induced resistance to RNA damage, ribonucleases (RNases) have great therapeutic potential as an alternative to the standard DNA-damaging chemotherapeutic agents, which develop resistance (2). It is known that the ribonucleolytic activities of RNases are responsible for their cytotoxicity. Furthermore, ONC has possible therapeutic implications in cancer therapy due to the induction of p53 independent apoptosis (3). Consequently, t-RNA damage cell death is a new and attractive cancer treatment modality, since a functional p53 is not affected by RNA damage. In this study, we found that ONC showed the enhanced radiation response of A549 human NSCLC *in vitro* and *in vivo*, likely due to reduced O_2 consumption (QO_2) and temporarily improved tumor blood flow, in addition to the previous mechanisms observed by various investigators (2-3).

2. METHODS AND MATERIALS

2.1. Tumors

Frozen A549 human NSCLC cells were purchased from the American Type Culture Collection (Rockville, MD), thawed, cultured, and grown *in vitro*. Cells were maintained at 37°C for the duration of the experiments. The cells grew in RPMI1640 medium supplemented with 15% FBS, 25mM HEPES buffer, 5mM l-glutamine, and antibiotics (15% FBS-RPMI1640)

2.2. Clonogenic survival for *in vitro* radiation response to ONC

ONC was supplied by the Alfacell Corporation (Bloomfield, NJ). The proper cell number (from 100 to 1×10^5) was plated into 6-well plates, incubated at 37°C for 5 hr for cell attachment, prior to ONC treatment. All cell cultures were exposed to various concentrations of ONC for 24 hr whereupon the adherent cells in 6 well-plates were irradiated using a ^{137}Cs irradiator at a dose rate of 1.35 Gy/min. After rinsing 6-well plates twice with RPMI1640 medium, 7 ml of 15% FBS-RPMI1640 were added to 6-well plates and allowed to grow for 8-9 days. To evaluate the effect of ONC on sublethal repair damage, cells plated in 6-well plates were treated with ONC and irradiated using a ^{137}Cs irradiator. The split dose schedule (first dose of radiation at 5 Gy and second dose at 7 Gy) was performed by incubation of cells at room temperature (to prevent the cells from moving through the cell cycle between the dose fractions) for 1, 2, and 4 hr between irradiation. Then, 6-well plates were rinsed twice with RPMI1640 medium and 7 ml of 15% FBS-RPMI1640 were added to 6-well plates and allowed to grow for 8-9 days. Cultures were fixed with 99.5% isopropyl alcohol, stained with 1% crystal violet, and counted. Colonies comprising >50 cells were scored as positive.

2.3. Measurement of oxygen consumption rate using a Clark-type electrode

The A549 tumor cells were cultured in the confluent levels, and the preparation of single cell suspensions of tumor cells were done with a scraper without adding a trypsin solution. For the procedure of QO_2 measurements, an YSI Biological Oxygen Monitor (Yellow Springs Instrument Co, Yellow Springs, OH) was utilized as described in detail in our previous report (4).

2.4. Measurement of apoptosis

A549 tumor cells (2.5×10^4 per chamber) were plated on either 2- or 4-chamber slides using 15% FBS-RPMI1640 medium (containing no phenol), incubated overnight prior to ONC treatment. Cells were exposed to ONC for 30 hr. Apoptosis measurement was performed by using the fluorescein (Roche applied Science, Germany) *in situ* cell death detection kit and a phase-fluorescent microscope.

2.5. Animals and tumors

The institution's guidelines for the care and use of laboratory animals were followed. Eight- to ten-week-old, female athymic NCR-nu/nu nude mice (purchased from the NCI, Bethesda, MD) bearing human tumor xenografts of A549 human NSCLC cells were utilized. 2×10^6 viable cells suspended in 50 μl of RPMI1640 medium were injected subcutaneously into the right thighs of mice. Experiments were carried out when the tumor volume was between 200 and 400 mm^3. Tumor volumes were calculated using the formula $V = 0.4 \times AB^2$, with A & B as the longer and shorter diameters of the tumor, respectively (5). For the *in vivo* ONC treatment, ONC was dissolved in sterile 0.9% NaCl solution before the experiments. The mice were given either an intravenous (i.v.) or intra-peritoneal (i.p.) injection of ONC at 2.5 mg/kg at a volume of 0.2 ml/ 20 g of body weight.

2.6. Animal preparation and data acquisition for MRI

All MRI images were acquired with a 4.7-T ^{UNITY}INOVA spectrometer (Varian Medical Systems, Palo Alto, CA). For the *in vivo* MRI studies, a catheter was inserted into the tail vein of the mice for injection of the contrast agent. Then mice were placed inside a ^1H linear polarized birdcage RF coil with an inner diameter of 5 cm. The animal was sedated with isoflurane through a nose cone. Its core temperature was monitored by a rectal thermistor and maintained at $37^\circ C$ by a flow of warm air. The electrocardiogram (ECG) was also taken by attaching a pair of sub-dermal needle electrodes. The protocol for T_1 map included a multi-slice various flip angle gradient echo pulse sequence with TE/TR of 2.1/200 msec, flip angles of $0 \sim 180^\circ$. Dynamic T_1-weighted spin echo pulse sequence was recorded with the following parameter settings: TE/TR of 8.5/200 msec, a time resolution of 12.8 sec, the total number of dynamic images of 300 per each slice, the total imaging time of 64 min, single acquisition, the same field of view as T_1 map protocol, and the same matrix as T_1 map protocol. A bolus injection of Gd-DTPA at a dosage of 0.5 mmol/kg (Omniscan^{TM}: Amersham Health, Buckinghamshire, UK) was followed by 10 pre-acquisitions.

2.7. Kinetic model

After securely connecting a catheter into the tail vein, Gd-DTPA was diluted to 50 mM, and infused as a bolus with administration time of ~12 sec to maintain heart rate in each animal. Image analysis of the dynamic contrast-enhanced data was based on the generalized kinetic model developed by Tofts and Kermode (6). This model predicts the changes in concentration of the contrast agent in the tissue due to its transfer from the plasma into the interstitial space and back.

$$C_t = D \cdot K^{trans} \sum_{i=1,2} a_i \frac{\exp\left(-K^{trans} t / V_e\right) - \exp\left(-m_i t\right)}{m_i - K^{trans} / V_e} \qquad \text{(Equation. 1)}$$

Where K^{trans} is the volume transfer constant between blood plasma and extravascular extracellular space (EES; unit is min^{-1}), V_e is the EES fractional volume that is the volume of EES per unit volume of tissue. The time course of signal enhancement during the 60 min after bolus injection was analyzed at pixel resolution using a nonlinear Levenberg-Marquardt least-square fitting algorithm.

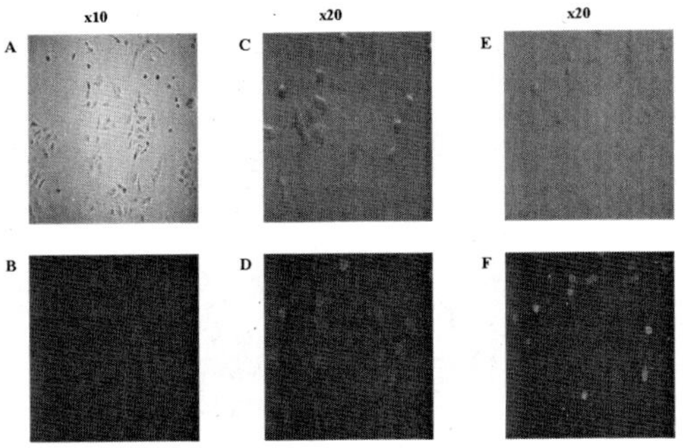

Figure 1. Apoptosis induced by ONC (A,B: 0, C,D: 5, E,F: 20 μg/ml-ONC). A, C and E are phase contrast images whereas B, D and F are fluorescent images.

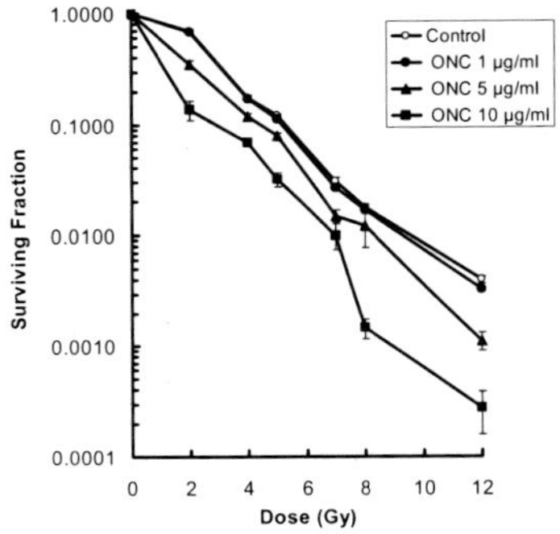

Figure 2. Radiation survival curves: increased radiation sensitivity observed after treatment with ONC.

3. RESULTS

ONC alone significantly inhibited the tumor growth of A549 tumors (IC_{50} = 1 μg/ml, data not shown). As shown in Figure 1, ONC significantly induced apoptosis. No occurrence of apoptosis in untreated A549 tumor cell lines was observed using a confocal microscope (photographs of A, C, E: phase-contrast images, photographs of B, D, F; fluorescent images). However, the ONC-induced apoptosis was found after 30 hrs incubation with ONC at 5 and 20 μg/ml (as shown in Fig. 1-D & F).

Using a clonogenic assay, we observed that ONC at 5-10 μg/ml could sensitize radiation response of A549 tumor cells *in vitro* by calculating the SER. SER was ~1.3 (P<.05) for ONC-5 (5 μg/ml) and ~1.6 (P<.05) for ONC-10 at a survival fraction of 0.01, as shown in Figure 2. However, ONC-1 did not show similar results. Interestingly, the initial shoulder of the ONC-treated radiation survival curves was significantly reduced by ONC. Thus, we proposed that ONC could inhibit sublethal damage repair (SLDR). The split dose experiment (5 Gy as the first dose, incubated at room temperature between irradiation, 7 Gy as the second dose) was performed to evaluate the effect of ONC on SLDR. The recovery factor (SF of the two split dose divided by SF of the one continuous dose) was ~2.7 to 3.1 for radiation alone. But, we found that ONC at 10μg/ml significantly reduced the recovery factor to 0.2 (data not shown).

In Figure 3, ONC-induced inhibition of QO_2 was significant in A549 tumor cell lines. The QO_2 values significantly decreased with longer exposure times to ONC. However, the QO_2 values stabilized after exposure to ONC between 15 and 20 min, and the reduced QO_2 at 10 min was ~40% of the control values. In Figure 4, ONC-2.5 (2.5 mg/kg) slightly retarded the A549 tumor growth by 3 days. However, ONC-10 significantly inhibited tumor growth by 15 days (data not shown). ONC-2.5 significantly improved radiation response in nude mice bearing A549 tumors. X-irradiation at 10 Gy with an i.v. injection of saline (as a control vehicle) significantly retarded tumor growth compared to the untreated control during the 60 day observation period. When ONC was i.v. administered 2 hr prior to x-irradiation, the tumor growth delay was >30 days compared to the radiation alone group. It was more than the additive effects of ONC and x-irradiation on the inhibition of tumor growth of A549 *in vivo*.

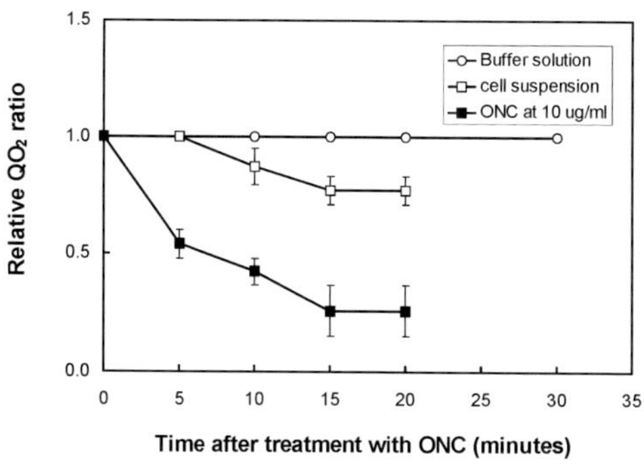

Figure 3. Cellular O_2 consumption (QO_2) inhibited by ONC.

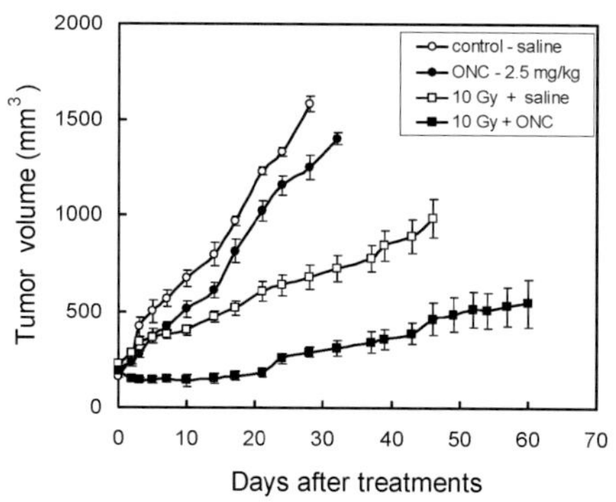

Figure 4. Growth delay assay after treatment with ONC and X-irradiation at 10 Gy.

For measurements of the plasma and tissue concentration curve, K^{trans} and V_e map, MRI data processing was done sequentially. The important and critical factor was obtaining the changes in the concentration of Gd-DTPA in the vena cava of an animal over the duration of the study of each animal. We tried to find the plasma concentration curve from each dynamic scan. This method provided the minimum error compared to the ideal plasma concentration curve. By DCE MRI using a diffusible MRI contrast agent of Gd-DPTA, we monitored the effect of ONC on tumor perfusion. K^{trans} and V_e were calculated from the changes in the MR signals using the physiological values (i.e., a_1, a_2, m_1, and m_2) and Equation 1. In the rim of tumors the signal enhancement after injection of Gd-DPTA rapidly increased, then rapidly decreased. However, in the core of tumors the signal enhancement slowly increased then slowly decreased after injection of Gd-DPTA (Fig. 5). There were observable increases in ROI with failing in fitting due to reduction in blood supply, particularly in the center of the tumor regions, although ONC did not alter the total tumor volume immediately after an injection of ONC. The K^{trans}

values at the rim of tumor regions were temporarily increased at 1.5 h post-injection of ONC (data not shown). However, the K^{trans} values (in rim + core) significantly reduced at 24 h post-treatment with ONC (Fig. 6).

4. DISCUSSION

The first goal of this study was to determine if ONC would enhance the radiation response of A549 human NSCLC cells *in vitro* and *in vivo*. We observed no apoptosis in A549 untreated control tumor cell lines using fluorescent images in Fig. 1-B, but ONC-induced apoptosis occurred at 30 h post-incubation with ONC (Fig. 1-D & F). In Figure 2, the control group's radiation survival curve had a shoulder, but this shoulder disappeared after treatment with ONC. Also, from the split dose experiment, we showed that ONC significantly inhibited the SLDR. With a short exposure to ONC, it inhibited cellular QO_2 at 10 min and reduced values to ~60% compared to the control values (Fig. 3). Since our x-irradiation protocol was done after 24 hr of exposure to ONC, this may have induced a near-complete inhibition of O_2 demand from the ONC-treated A549 tumor cells. Thus, ONC-induced radiation enhancement probably involved several cellular mechanisms including the inhibition of SLDR, inhibition of QO_2, and induction in apoptosis.

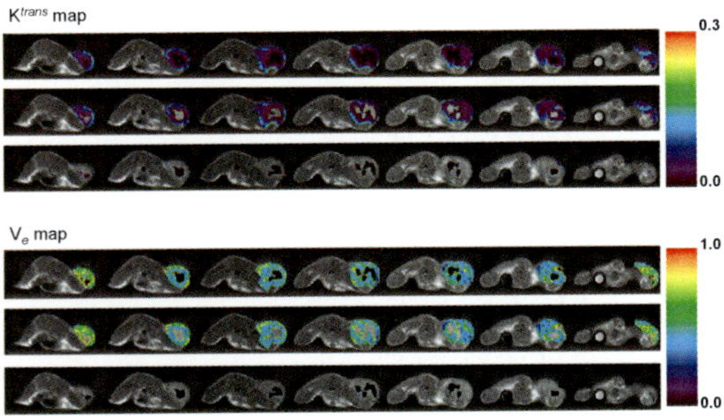

Figure 5. Permeability (K^{trans}) and interstitial space volume fraction (V_e) maps of A549 tumors.

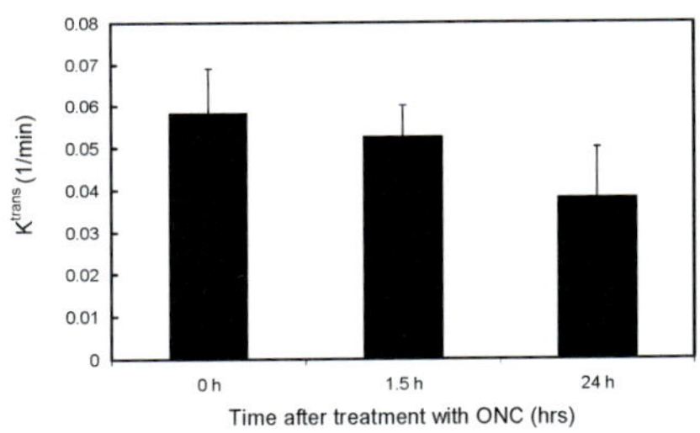

Figure 6. Changes in permeability (K^{trans}) of A549 tumors post-treatment with ONC.

Based on present *in vitro* studies, we performed the growth delay assay for A549 tumors in nude mice. To our knowledge, it is the first study of anti-tumoral effects of ONC on human NSCLC *in vitro* and *in vivo*. Furthermore, ONC enhanced radiation-induced tumoricidal effect on A549 human NSCLC xenografts in nude mice. A549 tumor cell lines are known to be $EFGR^{WT}$ and gefitinib resistant, but H1666 cell lines were $EFGR^{WT}$ and gefitinib sensitive *in vitro* (7). In general, gefitinib mono-therapy was not correlated with tumor expression of EFGR. Also, gefitinib did not increase apoptosis in A549 (7). Therefore, our findings have potential clinical implications that ONC could be a good candidate for the treatment of the gefitinib resistant NSCLC patients with $EFGR^{WT}$.

Using a non-invasive DCE-MRI technology, the K^{trans} values at the rim of tumor regions were temporarily increased at 1.5 h post-injection of ONC. This is in agreement with our previous observations using a laser Doppler flowmetry (5, 8). As described above, ONC significantly inhibited QO_2 in A549 tumor cell lines (Fig. 3). We concluded that the enhanced tumor oxygenation was mainly due to the reduction in QO_2 rather than an increase in tumor blood flow after treatment with ONC. Therefore, ONC significantly increased the radiation-induced tumor growth delay of A549 tumors *in vitro* and *in vivo*. This investigation suggests important potential clinical uses of ONC for the treatment of NSCLC cancer patients.

5. REFERENCES

1. Mikulski S.M., Costanzi J.J., Vogelzang N.J., McCachren S., Taub R.N., Chun H., Mittelman A., Panella T., Puccio C., Fine R., and Shogen K., Phase II trial of a single weekly intravenous dose of ranpirnase in patients with unresectable malignant mesothelioma, J. Clin. Oncol. 20, 274-281 (2001).
2. Rybak S.M. and Newton D.L., Natural and engineered cytotoxic ribonucleases: therapeutic potential, Experimental Cell Res. 253, 325-335 (1999).
3. Iordanov M.S., Ryabinina O.P., Wong J., Dinh T.H., Newton D.L., Rybak S.M., and Magun B.E., Molecular determinants of apoptosis induced by the cytotoxic ribonuclease onconase: evidence for cytotoxic mechanisms different from inhibition of protein synthesis, Cancer Res. 60, 1983-1994 (2000).
4. Lee I., Lee Y.H., Mikulski S.M., Lee J., and Shogen K., Enhanced cellular radiation sensitivity of androgen-independent human prostate tumor cells by onconase, Anticancer Res. 20, 1037-1040 (2000).
5. Lee I., Lee Y.H., Mikulski S.M., Lee J., Covone K., and Shogen K., Tumoricidal effects of onconase on various tumors, J. Surgical Oncology 73, 164-171 (2000).
6. Tofts P.S. and Kermode A.G., Measurement of the blood-brain barrier permeability and leakage space using dynamic MR imaging. 1. Fundamental concepts, MRM 17, 357-367 (1991).
7. Tracy S., Mukohara T., Hansen M, Meyerson M., Johnson B.E., and Jaenne P.A., Gefitinib induces apoptosis in $EFGR^{L858R}$ non-small-cell lung cancer line H3255, Cancer Res. 64, 7241-7244 (2004).
8. Lee I., Lee Y.H., Mikulski S.M., Shogen K., Effect of onconase $\pm$ tamoxifen on AsPC-1 human pancreatic tumors in nude mice, Adv. Exp. Med. Biol. 530, 187-196 (2003).

EFFECT OF pH AND IMIDAZOLE ON PROTEIN C PURIFICATION FROM COHN FRACTION IV-1 BY IMAC

James J. Lee, Duane F. Bruley and Kyung A. Kang *

Abstract: Cohn Fraction IV-1 (CFIV-1) is a by-product (often discarded) of a plasma fractionation process. It retains 90% of Protein C (PC) of plasma but contains several coagulants structurally homologous to PC. Of these coagulants, Factor II (FII) has the longest half-life (12 times of PC) and largest quantity (9 times of PC) in CFIV-1. Current purification process for PC is by immunoaffinity chromatography using monoclonal antibodies, which is very expensive. Immobilized metal affinity chromatography (IMAC) is an inexpensive process that uses metal ions to adsorb proteins *via* their surface histidines. Affinity of PC to the metal ions in IMAC is higher than that of FII because PC has 15 surface histidines and FII has 5. Two important factors in an IMAC process are pH and imidazole concentration. PH controls protonation of histidine, and imidazole, a histidine analog, competitively reacts with metal ions. The effects of pH and imidazole on adsorption and elution of PC and FII during IMAC process were studied. The effect of pH on PC and FII adsorption was similar within the range of 6.0 and 8.0. At concentrations below 15 mM imidazole, little PC or FII eluted. At 15 and 20 mM imidazole 2.5% of PC was eluted, while 20-30% of FII was eluted.

1. INTRODUCTION

Protein C (PC) has the functions of an anticoagulant, antithrombotic, anti-inflammatory,[1,2] and antiapoptotic[3,4] in plasma. PC, therefore, can be an effective therapeutic for patients with PC deficiency, thrombo-embolisms, advanced sepsis, and stroke.[5] Our study is to develop a process that can purify PC by an inexpensive means, specifically, by immobilized metal affinity chromatography (IMAC).

Sources of PC are transgenic animal milk, mammalian cells, and blood plasma. Cohn Fraction IV-1 (CFIV-1) has usually been the waste in the human serum albumin

* James J. Lee and Kyung A. Kang, Department of Chemical Engineering, University of Louisville, Louisville, KY 40292. Duane F. Bruley, Department of Chemical and Biochemical Engineering, University of Maryland Baltimore County, Baltimore, MD, 21250 and Synthesizer Inc., Columbia, MD.

(HSA) separation process during the plasma fractionation, and, therefore, is a very cost effective PC source.[14] In CFIV-1, there is approximately 100 μg-PC/g-pastes. CFIV-1, however, has several coagulant proteins whose amino acid sequences have 60-71% homology to that of PC.[6] These proteins are Factors II (prothrombin), VII, IX, and X.[6, 7] Among these coagulant proteins, FII is the most problematic because its amount in CFIV-1 is about 12 times greater than that of PC,[8] and has an average half-life of 2-5 days[9] while that of PC is only ten hours.[10] FX has a half-life of 32-48 hours,[9] but in CFIV-1, its amount in plasma is only about 3 times greater than that of PC.[8] FIX and FVII are of little concern because their amounts in CFIV-1 are less than that of PC.[8, 9] Thus, our study focus is to remove FII from the PC product, which may increase thrombotic risk if not adequately separated from PC.[11]

Immunoaffinity chromatography (IAC) is the current method used for purifying PC from blood plasma. This method is highly specific for the targeted molecule but expensive. With a well-designed process, IMAC may be an inexpensive and alternative method to IAC. In our preliminary studies, Wu et al. demonstrated that IMAC could separate PC from FII.[12, 13, 14] In IMAC, the number of accessible histidines is a critical factor for the separation process because they are the receptor for the metal ions on the gel matrix.[15, 16] Based on our analysis using the 3-D protein structures, PC differs from FII in the number of surface histidine residues that can react with the immobilized Cu^{2+} ion: PC has 18 histidine residues and 15 residues are exposed to the surface, while FII has 5 histidine residues, with 4 of them on the surface.[17, 18]

Our study goal is to optimize the IMAC process for PC separation from FII in CFIV-1 by selecting appropriate operating conditions that affect protein adsorption and desorption based on the preliminary study results obtained by Wu et al. The affinity between the protein and the immobilized metal ion depends upon the liquid phase pH, which controls the level of protonation.[19] The histidine residues of a protein become more protonated as pH is decreased, and thus, decrease their affinity to the metal ion.[19] Our study aim is to determine the pH that would maximize PC adsorption and minimize FII adsorption to the metal ions on the gel matrix. In IMAC, imidazole is frequently used to elute the product from the matrix because it reacts to the metal ions competitively against histidine. Wu et al., have used a step-wise elution with 2 mM imidazole followed by 15 mM imidazole to separate a PC from FII.[12] Our second study aim is, therefore, to study the effect of imidazole concentration on PC and FII elution, to achieve maximum separation of PC from FII.

2. INSTRUMENTS, MATERIALS, AND METHODS

CFIV-1 paste was generously donated from the American Red Cross (Rockville, MD). CFIV-1 paste was pretreated by, first, dissolving it in 20 mM sodium citrate buffer at pH 6.0, at a ratio of 1 g-paste/30 ml-buffer for 5 hours at 4°C. The dissolved paste was then centrifuged at 4000 rpm for 30 minutes at 4°C using a centrifuge (Marathon 3200R; Fisher Scientific; Chicago, IL).

For IMAC processes, the Chelating Sepharose Fast Flow™ matrix with iminodiacetic acid (IDA) (Pharmacia Biotech; Piscataway, NJ) was prepared as described by Nandakumar, et al.[20] Cupric Sulfate Pentahydrate (Sigma; St. Louis, MO) was used as the source of Cu^{2+}. The matrix was equilibrated with 20 mM sodium phosphate buffer (0.5 M NaCl). 100 μl Cu^{2+}-IDA matrix was reacted with (Innovative Research; Southfield, MI) 200 μl of sample at room temperature (22-24°C). The vial was gently vortexed for 2 seconds to suspend the matrix to facilitate mixing, and then was placed on a mixer (Clay Adams® Nutator; Becton-Dickinson; Franklin Lakes, NJ) with gentle agitation for ten minutes. The suspensions were then centrifuged at 4000 rpm at 4°C for one minute, and the protein in the supernatant was quantified by ELISA.

For the study of protein elution by imidazole, the protein adsorbed matrix was mixed with 400 μl of the equilibrium buffer for 10 minutes to remove any loosely adsorbed proteins. The suspension was centrifuged and the supernatant was removed. Then 400 μl of the elution buffer with imidazole at a predetermined concentration was applied to the matrix and reacted for 10 minutes, and then, the suspension was centrifuged. The protein in the supernatant was quantified by ELISA.

ELISA of PC and FII were performed as follows: EIA/RIA 96-well flat-bottom plates (Corning, NY) were first coated with Rabbit anti-PC IgG (Sigma; St. Louis, MO), or Rabbit anti-FII IgG for (Biomeda; Foster City, CA). The plate was washed and blocked with BSA for 90 minutes. After blocking, the plate was washed and samples were applied and incubated for 90 minutes. Then the plate was washed and incubated for 90 minutes with Goat anti-human PC IgG (American Diagnostica Inc.; Hauppauge, NY), or Mouse anti-FII (Enzyme Research; Southbend, IN). Next, the plate was washed and incubated for 20 minutes with horseradish peroxidase-conjugated to anti-goat IgG rabbit antibody, or horseradish peroxidase-conjugated to anti-mouse IgG goat antibody (Sigma). Then o-phenylenediamine dichloride (OPD; Sigma) was added for color development. The optical density was measured at 450 nm using and ELISA plate reader (Bio-Rad; Hercules, CA).

3. RESULT AND DISCUSSION

3.1. Effect of pH on Adsorption

In IMAC, as previously stated, pH affects protein adsorption by protonating or deprotonating the electron donating site on the histidine residue. PC has a number of exposed histidine residues different from FII. Thus, it was expected that the adsorption level of PC on the Cu^{2+}-IDA matrix at a pH was different from that of FII. Here, the effect of pH on the adsorption of PC and FII to the matrix was studied at a pH range between 6.0 and 8.0. This range was selected because it is close to the physiological pH, and is in the range that favors histidine adsorption in IMAC.[19]

20 μg of PC or FII in 200 μl of equilibration buffer at a pH 6.0, 7.0, or 8.0 was reacted with 100 μl of the matrix for 10 minutes at a room temperature. The maximum protein adsorption capacity of 100 μl of the Cu^{2+}-IDA matrix is expected to be between 10 and 11 mg.[21] Then, the matrix was separated by centrifugation, and the PC/FII in the liquid phase amount was quantified by ELISA. This same experiment was repeated for 40 μg proteins.

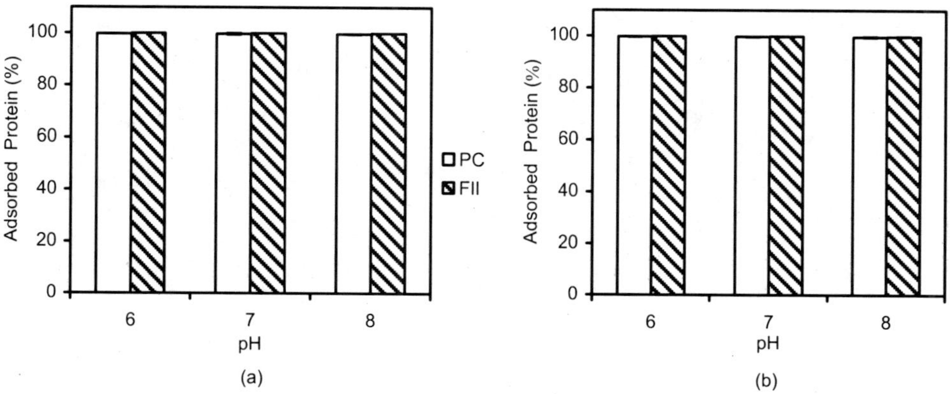

Figure 1. The amount of PC or FII adsorbed at various pHs: when (a) 20 μg and (b) 40 μg protein was reacted with 100 μl Cu^{2+}-IDA matrix for 10 minutes.

Figure 1 shows the amount of PC and FII adsorbed to the matrix at various pHs. The amount of the reacted protein [20 μg (Fig. 1a) or 40 μg (Fig. 1b)] did not influence the adsorption level, and for all cases, most of the added protein was adsorbed to the matrix. There were little differences in the adsorption level between PC and FII at the pH ranges tested; indicating that pH alone may not change the adsorption level of these two proteins.

3.2. Effect of Imidazole on the Elution

Because PC and FII have a different number of exposed histidine residues, PC was expected to be eluted the best at an imidazole concentration different from that FII is eluted. Therefore, the elution of PC/FII was studied at an imidazole concentration range between 2 and 20 mM. This range was selected based on preliminary study results by Wu et al.[12, 13, 14]

Either 20 or 40 μg of PC/FII was adsorbed in 100 μl of Cu^{2+}-IDA matrix at pH 6.0. Then 400 μl of imidazole solution at a concentration between 2 and 20 mM at pH 6.0 was mixed with the PC/FII adsorbed matrix, and reacted for 10 minutes. After the sample was centrifuged, PC/FII in the supernatant was quantified by ELISA.

Figure 2 shows the amount of PC and FII eluted by imidazole at various concentrations. For PC, less than 2.5% of the total adsorbed was eluted at imidazole concentrations between 2 and 20 mM. For FII, at the imidazole concentrations lower than 15 mM, very little amount was eluted. At 15 and 20 mM, 20~30% of the total FII adsorbed was eluted (with slightly higher for 15 mM imidazole) for both matrices with 20 and 40 μg. Although the elution of FII was significant at 15 and 20 mM, 70 to 80% of FII still remained in the matrix. The amount of PC/FII adsorbed to the matrix (e.g. 20 or 40 μg/100 μl-matrix) did not significantly affect the level PC/FII eluted at various imidazole concentrations. PC can be separated from FII by eluting FII with imidazole concentrations at 15 and 20 mM.

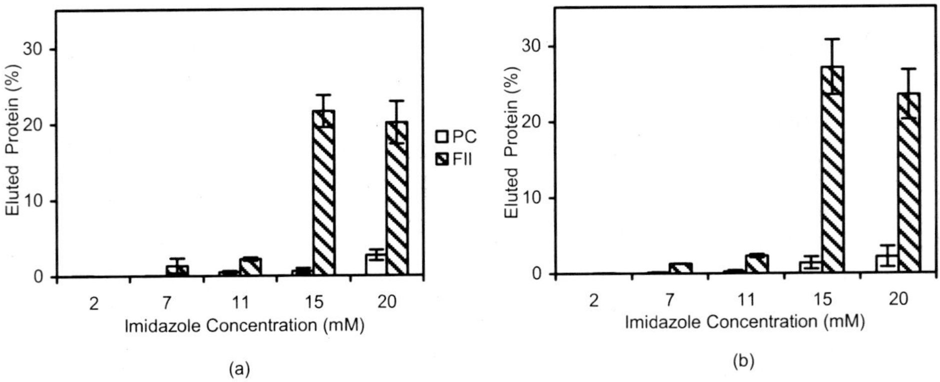

Figure 2. The amount of PC/FII eluted from the matrix with (a) 20 and (b) 40 μg of protein by imidazole at various concentrations.

4. CONCLUSIONS

The effect of pH on PC and FII adsorption to the Cu^{2+}-IDA matrix were studied to determine a pH that can maximize PC adsorption and minimize FII adsorption. The affinity of PC and FII to the chelated matrix was high to adsorb at least 400-μg protein/ml of matrix with very little difference between the two proteins. PH in the range between 6.0 and 8.0 does not appear to be a good operational condition for separating PC from FII. Since pH does not significantly affect protein adsorption, pH 6.0 may be used for the benefit of minimizing the possible protease activity in the source material. PC may be separated from FII during the elution process at an imidazole concentration of 15~20 mM.

5. FUTURE STUDY

Imidazole at a lower concentration will be added in the adsorption step to determine if it may provide more specificity for adsorption of PC or FII to the matrix. The elution of PC and FII at the imidazole concentrations higher than 20 mM will be examined to improve FII elution with minimal PC elution.

6. AKNOWLEDGEMENTS

The authors thank the American Red Cross (Rockville, MD) for providing Cohn Fraction IV-1.

7. REFERENCES

1. C. T. Esmon, The Anticoagulant and Anti-Inflammatory Roles of the Protein C Anticoagulant Pathway, *J. Autoimmun.* **15,** 113-116 (2000).
2. C. T. Esmon, Protein C anticoagulant pathway and its role in controlling microvascular thrombosis and inflammation, *Crit. Care Med.* **29**(7)**,** 48-51 (2001).
3. D. E. Joyce, L. Gelbert, A. Ciaccia, B. DeHoff and B. W. Grinnell, Gene expression profile of antithrombotic protein C defines new mechanisms modulating inflammation and apoptosis. *Journal of Biol. Chem.*, **276** 11199-11203 (2001).
4. T. Cheng, D. Liu, J. H. Griffin, J. A Fernandez, F. Castellino, E. D. Rosen, K. Fukudome and B. V. Zlokovic, Activated protein C blocks p53-mediated apoptosis in ischemic human brain endothelium and its neuroprotective, *Nat. Med.*, **9** (3) 338-342 (2003).
5. D. Liu, T. Cheng, H. Guo, J. A. Fernandez, J. H. Griffin, X. Song and B. V. Zlokovic, Tissue plasminogen activator neurovascular toxicity is controlled by activated protein C, *Nat. Med.*, **10** (12) 1379-1383 (2004).
6. R. M. Bertina, *Protein C and Related Proteins; Biochemical and Clinical Aspects*, (Churchill Livingstone, New York, 1998), pp. 1-54.
7. D. F. Bruley and W. N. Drohan, *Advances in Applied Biotechnology Series; Protein C and Related Anticoagulants*, **11** (Gulf Publications, Houston, TX 1990).
8. S. Rezania, D. G. Ahn and K. A. Kang, Separation of Protein C from Cohn Fraction IV-1 by Mini-Antibody, Proceedings of the 33rd Annual ISOTT meeting 2005 [In Review].
9. L. H. Edmunds and E. W. Salzman, Hemostatic Problems, Transfusion Therapy, and Cardiopulmonary Bypass in Surgical Patients, *Hemostasis and Thrombosis*, 3rd Ed., edited by R. W. Colman, J. Hirsh, V. J. Marder and E. W. Salzman, (J. B. Lippincott Co, Philadelphia, 1994), p. 958.
10. G. J. Broze Jr. and J. P. Miletich, Biochemistry and Physiology of Protein C, Protein S, and Thrombomodulin, *Hemostasis and Thrombosis*, 3rd Ed., edited by R. W. Colman, J. Hirsh, V. J. Marder and E. W. Salzman, (J. B. Lippincott Co., Philadelphia, 1994), p. 262.

11. A. S. Wolberg, D. M. Monroe, H. R. Roberts and M. Hoffman, Elevated prothrombin results in clots with and altered fiber structure: a possible mechanism of the increased thrombotic risk, *Blood*, **101,** 3008-3013 (2003).
12. H. Wu and D. F. Bruley, Homologous human blood protein separation using immobilized metal affinity chromatography: protein C separation from prothrombin with application to the separation of factor IX and prothrombin, *Biotechnol. Progr.* **15,** 928 (1999).
13. H. Wu, and D. F. Bruley, Chelator, metal ion and buffer studies for protein C separation, *Comp. Biochem. Physiol. Part A* **132,** 213-220 (2002).
14. H. Wu, D. F. Bruley, K. A. Kang, Protein C Separation form human plasma Cohn fraction IV-1 using immobilized metal affinity chromatography, *Adv. Exp. Med. & Bio.: Oxy. Trans. to Tis. XX,* **454,** (Plenum Press, New York, 1998), pp. 697-704.
15. M. Zachariou and M. T. W. Hearn, Characterization by Potentiometric Procedures of the Acid-Base and Metal Binding Properties of two New Classes of Immobilized Metal Ion Affinity Adsorbents Developed for Protein Purification, *Journal of Analytical Chemistry*, **69**, 813-822 (1997).
16. E. S. Hemdan, Y. J. Zhao, E. Sulkowski and J. Porath, Surface topography of hisitdine residues: A facile probe by immobilized metal ion affinity chromatography, *Proceedings of the National Academy of Science USA*, **86** (1989) 1811-1815.
17. J. J. Lee, E. E. Thiessen and D. F. Bruley, Protein C Production; Metal Ion/Protein Interfacial Interaction in Immobilized Metal Affinity Chromatography, *Adv. Exp. Med. & Bio.: Oxy. Trans. to Tis. XXVI*, **566**, (Springer-Verlag, New York, 2005).
18. E. E. Thiessen, and D. F. Bruley, Theoretical Studies of IMAC Interfacial Phenomena for the Purification of Protein C, *Adv. Exp. Med. & Bio.: Oxy. Trans. to Tis. XXV*, **540**, (Kluwer Academic/Plenum Press, New York, 2003), pp 183-190.
19. E. K .M. Ueda, P. W. Goutt and L. Moranti., Rev. of "Current and prospective application of metal ion-protein binding," *J. of Chromatography A* **988**, 1-23 (2003).
20. R. Nandakumar, H. Afshari and D. F. Bruley, Analysis of Equilibrium Adsorption Isotherms for Human Protei C Purification by Immobilized Metal Affinity Chromatography, *Adv. Exp. Med. & Bio.: Oxy. Trans. to Tis. XXV*, **540**, (Kluwer Academic/Plenum Press, New York, 2003), pp. 191-199.
21. T. W. Hutchens, T. T. Yip and J. Porath, Protein Interaction with Immobilized Ligands: Quantitative Analyses of Equilibrium Partition Data and Comparison with Analytical Chromatographic Approaches Using Immobilized Metal Affinity Chromatography, Analytical Biochemistry **170,** 168-182 (1988).

PREDICTING MELANOMA METASTATIC POTENTIAL BY OPTICAL AND MAGNETIC RESONANCE IMAGING

Lin Z.J. Li, Rong Zhou, Tuoxiu Zhong, Lily Moon, Eun Ju Kim, Hui Qiao, Stephen Pickup, Mary J. Hendrix, Dennis Leeper, Britton Chance, Jerry D. Glickson*

Abstract: Accurate prediction of tumor metastatic potential would be helpful in treatment planning and in the design of agents that modify the tumor phenotype. We report that three methods that are potentially transferable to the clinic – dynamic contrast enhanced MRI (DCE MRI), $T_{1\rho}$-weighted imaging and low temperature fluorescence imaging (that could be performed on biopsy specimens) – distinguished between relatively indolent (A375P) and aggressive (C8161) metastatic human melanoma xenografts in nude mice, whereas T_1 and T_2 relaxation time measurements did not. DCE MRI data analyzed by the BOLus Enhanced Relaxation Overview (BOLERO) method in conjunction with concurrent measurements of the arterial input function yielded a blood transfer rate constant (K_{trans}) which measures perfusion/permeability, that was significantly higher in the core of the indolent tumor than in the core of the aggressive tumor. Histological staining indicated that aggressive tumors had more blood vascular structure but fewer functional vascular structure than indolent tumors. Indolent tumors exhibited $T_{1\rho}$ values that were significantly higher than those of aggressive tumors at spin-locking frequencies >500Hz. The mitochondrial redox ratio, Fp/(Fp+NADH), where Fp and NADH are the fluorescence of oxidized flavoproteins and reduced pyridine nucleotides, respectively, of aggressive tumors was much higher (more oxidized) than that of indolent tumors and often showed a bimodal distribution with an oxidized core and a reduced rim. These differences observed between these two types of tumors, one indolent and one aggressive, if generalizable, would be very valuable in predicting human melanoma metastatic potential.

* Lin Z.J. Li, Rong Zhou, Eun Ju Kim, Hui Qiao, Stephen Pickup, Jerry D. Glickson (glickson@mail.med.upenn.edu, ph 215-898-1805, fax 215-573-2113), Molecular Imaging Laboratory, Department of Radiology, Philadelphia, Pennsylvania 19104. Tuoxiu Zhong, Lily Moon, Britton Chance, Department of Biochemistry & Biophysics, Johnson Research Foundation, University of Pennsylvania, Philadelphia, Pennsylvania 19104. Mary J. Hendrix, Children's Memorial Research Center, Northwestern University, Evanston, IL 60208. Dennis Leeper, Department of Radiation Oncology, Thomas Jefferson University, Philadelphia, PA19107.

1. INTRODUCTION

Accurate prediction of tumor metastatic potential by imaging methods during or soon after diagnosis of cancer would provide valuable information for guiding the type and aggressiveness of therapeutic procedures. Such methods could also provide valuable insight about the characteristics of the disease that could be utilized in the design of novel therapeutic strategies. In addition, imaging can non-invasively provide information about tumor metabolic and physiological heterogeneity.

The Hendrix laboratory has assembled a collection of human melanoma cell lines for which clinical data were available on the aggressiveness of the disease in patients. In three-dimensional cell culture, the aggressive melanoma cells express endothelial cell-associated genes and form vasculogenic-like networks that are laminin-positive, which correspond to the laminin-rich patterned networks observed in patient tumors, correlating with a poor prognosis.[1-3] This intriguing finding was called vasculogenic mimicry, and since its original description in aggressive melanomas, it has been independently confirmed in a variety of other tumor types.[2] In addition, recent studies have shown that melanoma vasculogenic mimicry corresponds to an extravascular perfusion pathway within aggressive tumors which communicates with the host vasculature and permits the perfusion of fluid via the anti-coagulation interaction(s) of Tissue Factor-positive melanoma cells with Tissue Factor Pathway Inhibitor(s).[4,5] We have evaluated the formation of melanoma tumors by dynamic contrast enhanced (DCE) MRI imaging and examination of pathology specimens. However, our principal objective was to evaluate imaging methods that could be applied in the clinic to distinguish between aggressive and indolent forms of this disease in the human patient. For this purpose we have established xenografts in athymic nude mice of the most aggressive human melanoma cell line (C8161) and the most indolent cell line (A375P). It has been previously demonstrated that the aggressive subline metastasizes within three weeks after subcutaneous inoculation while the indolent subline shows little sign of metastasis after implantation in SCID mice.[6,7]

As potential diagnostic imaging methods, we have chosen low temperature optical imaging or "redox scanning",[8] which can be clinically implemented with biopsy specimens and two clinically applicable noninvasive magnetic resonance imaging (MRI) techniques—dynamic contrast enhanced (DCE) MRI utilizing Gd-DTPA as the contrast agent and $T_{1\rho}$–MRI. Low temperature optical imaging provides very high spatial resolution 3D images (~80-100 μm in-plane and ~10 μm slice thickness) of tumor heterogeneity and monitors the redox status of mitochondria by measuring relative levels of reduced pyridine nucleotides (NADH) and oxidized flavoproteins (Fp), the latter originating from pyruvate dehydrogenase and α-ketoglutarate dehydrogenase. Redox imaging also monitors hemoglobin saturation and, hence, the state of tumor vascular (chiefly venous) oxygenation. The MRI methods provide moderate spatial resolution (~250 to 500 μm in-plane; 1 mm slice thickness). Dynamic contrast enhancement is sensitive to tumor perfusion, vascular permeability and extracellular volume fraction, whereas $T_{1\rho}$-weighted MRI is sensitive to macromolecular concentration and protein-water hydrogen exchange rates.

2. METHODS

2.1. Animal Models

Aggressive (C8161)[7] and indolent (A375P)[9] human melanoma cells were grown in RPMI medium supplemented with 10% fetal bovine serum. Melanoma cells ($2x10^6$) were subcutaneously implanted on the back shoulders or in the thighs of 7-9 week old male athymic nude mice (NCr-*nu/nu* 01B74) obtained from National Cancer Institute, US. After 1-2 months, the mice were examined by MRI and then sacrificed for redox imaging. The tumor volumes ranged approximately 0.2-0.4 cm^3, for C8161 and A375P xenografts, respectively, at the time of the experiments. For histological examination, excised tumors were fixed in formalin, embedded in paraffin, sectioned into 5 μm sections, fixed and stained with Factor VIII (DAKO, Carpinteria, CA) for detection of vascular endothelial cells and then counterstained with hematoxylin.

2.2. Rapid Tumor Freezing Procedure

Both types of the tumor bearing mice were anaesthetized by intraperitoneal injection of ketamine (100mg/kg) and acepromazine (10mg/kg), quickly frozen by immersion into liquid nitrogen (-196°C), and the subcutaneous tumors were not observed to exhibit cracks. Tumors were then surgically excised by saw, embedded in ethanol: glycerol: water mixture (1:3:6 by volume; freezing point, -30°C), and mounted for imaging.

2.3. Imaging of Tumors by Redox Scanning

The frozen tumor sample was milled flat and imaged every ~500 μm from the top surface to the bottom of the tumor. The light guide (fused silica, 50 μm core diameter) was stepped across the tissue surface at a distance of about 70 μm from the tissue surface. The excitation filter and the emission filter for detecting the fluorescent signals of each substance were designed on the basis of the absorption and emission spectra of each fluorophore and the emission spectrum of the mercury arc lamp. The fluorescent signals of Fp (filters: excitation 440DF20 (center wavelength 440 nm and band-pass width 20 nm); emission 520DF40), NADH (filters: excitation 365HT25, emission 455DF70), as well as the absorption of hemoglobin (Hb) at wavelength 546 nm (filter at emission: 546nm) and wavelength 578 nm (filter at emission: 578DF10) were imaged at each tumor depth. Scanning was performed in steps of 128×128 covering 1.0×1.0 cm^2 or 1.3×1.3 cm^2 of surface area with an in-plane resolution of 80 μm or 100 μm.

Using MATLAB™ data analysis programs developed in the Chance laboratory, the redox ratios Fp/(Fp+NADH) and NADH/(Fp+NADH) were calculated from the digitized signals of Fp and NADH, and the relative hemoglobin oxygenation map was calculated from the ratio of the signal intensity of Hb at 578 nm to its signal intensity at 546 nm.

2.4. DCE MRI of Melanoma Xenografts *in vivo*

Details of the procedure for performance of the DCE experiments and the methods for data analysis have been described in recent publications from the Zhou/Glickson laboratory.[10] Briefly, studies were performed on melanoma xenografts implanted on the backs of nude mice in order to permit simultaneous imaging of the left-ventricle

(ECG- gated at end-diastole) and the tumor in the same field of view. Studies were performed with a home-built birdcage coil on a Varian ANOVA spectrometer interfaced to a 4.7 T 30 cm bore magnet utilizing a 12 cm gradient insert capable of generating magnetic field gradients of 25 G/cm. Anesthesia with 1% isofluorane was administered through a nose cone (flow rate, 0.8 L/min) placed over the animal's snout. Tumors (8-10 mm in diameter) were imaged with a T_1-weighted pulse sequence at a rate of about 2 second/image, simultaneously with the acquisition of the arterial input function of Gd-DTPA, which was administered through a tail-vein catheter extending outside the magnet. An inversion-recovery pulse sequence was used to quantify T_1 of arterial blood and the tumor tissue.

Data analysis was accomplished with the BOLERO algorithm[11] that takes into account the rate of water exchange across the plasma membrane of the tumor cell relative to the difference in water relaxation rates in the intra- and extracellular compartments. Data modeling was performed using the average signal intensities of DCE images in two regions of interests, the tumor core and rim as described previously in Ref 10. Two-limiting conditions were evaluated: 1) the transcytolemmal water exchange rate was infinitely rapid compared to the difference in water relaxation rates between intra- and extracellular compartments (called the fast-exchange limit or FXL), and 2) the intra-extracellular water relaxation rate difference was comparable to the transcytolemmal water exchange rate when the large bolus of Gd-DTPA first reached the tumor (called the fast exchange regime or FXR), but was otherwise assumed to be infinitesimally small relative to the transcytolemmal water exchange rate. The first method that is commonly used by most workers in the field is called the "FXL-constrained method" and yields two parameters – the blood transfer rate constant K_{trans} and the volume fraction of the extravascular extracellular space in tissue v_e; the second method (called the "FXR-permitted method") that was recently introduced by Springer's laboratory[11] utilizes the FXR equations to analyze data that satisfy this limit and otherwise utilizes the FXL equations. Data analysis with the FXR equations yields three fitting parameters – K_{trans}, v_e, and the intracellular life time τ_i ($1/\tau_i$ is the transcytolemmal water exchange rate). The difference between these assumptions is subtle, but the choice of modeling methods can have a profound effect on the values of parameters that are calculated.

2.5. MR T_2 and $T_{1\rho}$-weighted Imaging of Melanoma Xenografts

The water $T_{1\rho}$ relaxation time characterizes the relaxation process of hydrogen nuclear spins in a coordinate system that is rotating at the proton Larmor frequency. The $T_{1\rho}$-weighted imaging pulse sequences utilize magnetization prepared by a fast spin-echo sequence as described by Reddy *et al.*[12,13] Briefly, a non-selective 90^0 hard pulse rotates proton spin magnetization into the transverse plane. An on-resonance spin-locking radiofrequency pulse (SLP) then locks the spins in the transverse plane perpendicular to the static magnetic field B_0. The SLP pulse has a duration of TSL (time of spin locking), during which time the spin magnetization relaxes with a time constant $T_{1\rho}$. Six values of TSL have been used in the range of 16-125 ms. Then another 90^0 hard pulse rotates the magnetization back to the z-direction, and data acquisition ensues with a conventional fast spin-echo pulse sequence (TR 1 sec, echo spacing 6 msec, matrix 128×128, FOV 3~4 cm, slice thickness 2 mm, NEX = 2). Values of $T_{1\rho}$ depend on the strength of the magnetic field B_1, which defines a spin-locking frequency, SLF = γB_1 (where γ is the gyromagnetic ratio of nuclear spins). We performed $T_{1\rho}$ dispersion measurement by changing B_1 from 300-4000 Hz. Imaging at each spin-locking frequency required about 2 minutes.

The signal (S) detected by MR $T_{1\rho}$-weighted imaging is an exponential defined by $S \propto \exp[-TSL/T_{1\rho}]$. By fitting the logarithm of S as a function of TSL utilizing customized software written in Interactive Data Language, we determined $T_{1\rho}$ for every image pixel at a specific frequency of the spin-locking pulse. We also obtained the mean value of $T_{1\rho}$ by fitting the average signal values in a region of interest drawn across a slice section of the tumor.

The T_2-weighted images were acquired with a conventional spin-echo sequence with TR 500 ms and 6 TEs varying from 12 to 62 ms. Values of T_2 were determined by fitting the T_2-weighted imaging signal intensity to the exponential function, $\exp[-TE/T_2]$.

3. RESULTS AND DISCUSSION

3.1. Optical Imaging

The ratio of Fp and NADH signals can approximate the oxidation-reduction ratio of the mitochondrial steady-state metabolism because: 1) the two signals represent the major fluorochromes in the mitochondrial matrix, and 2) the two chemicals are near oxidation-reduction equilibrium.[14] Therefore, the Fp redox ratio, Fp/(Fp+NADH), is a sensitive metabolic index of mitochondrial respiration of the cell.

A number of distinct respiratory states can be distinguished on the basis of the redox ratio.[15,16] State 1 corresponds to adequate oxygen and low levels of ADP and endogeneous substrates. This condition is indicative of low levels of oxidative metabolism accompanied by high NADH and low Fp, i.e., low Fp redox ratio. State 2, in which mitochondria are starved of substrate but have adequate oxygen and ADP, is characterized by very high levels of Fp and low levels of NADH, i.e., high Fp redox ratio. State 3, which corresponds to high levels of oxygen, ADP, and substrate and, hence, high levels of oxidative metabolism is also accompanied by a high Fp redox ratio, but not as high as State 2, and low NADH, but not as low as State 2. State 4 exhibits low mitochondrial activity with low ADP levels but with adequate supplies of substrate. This state exhibits high NADH redox ratio (NADH/(Fp+NADH)) but low Fp redox ratio.

Shown below are typical images of redox ratios of two sequential tissue slices of an aggressive C8161 xenograft versus an indolent A375P xenograft (Figure 1). The tissue optical penetration depth = 10 μm, slice depth 600 and 1200 μm beneath the skin. Distinct tumor heterogeneity in the spatial distribution of redox ratio Fp/(Fp+NADH) was observed in the 1200 μm depth image of the aggressive tumor. In the 600 μm tissue slice, on average, the aggressive C8161 xenograft was at least 2 times more oxidized (mean=0.66, STD=0.16) than the indolent A375P xenograft (mean=0.30, STD=0.12) at the same slice depth. We attribute this to State 2 metabolism in the aggressive tumor. State 3 might also yield similar redox ratios, but the low level of perfusion/permeability of this region of the tumor (see section 3.2) argues in favor of State 2. However, both tumors exhibited similar and relatively uniform levels of vascular oxygenation as indicated by their respective Hb ratio images (Figure 2).

An interesting bi-modal spatial distribution of redox ratios was observed in the deeper slices of the aggressive tumor, which was significantly more oxidized in the central core of the tumor (mean 0.70, STD 0.14) than in the outer rim (mean 0.15, STD 0.10) of the tumor (see the Fp redox ratio histogram of C8161 in Figure 1). We attribute this to State 2 respiration in the core and State 4 in rim (see section 3.2) of the aggressive tumor. This again reflects a relatively poorly perfused and metabolically less active core in the aggressive tumor, which is likely to generate VEGF that could contribute to a higher metastatic potential.

In comparison, the redox ratios of the mouse's leg muscle demonstrated a different distribution pattern than melanomas, less heterogeneity as can be seen from the histogram (Figure 3). Its average Fp redox ratio (mean 0.34) was less than that of an A375P tumor (mean 0.45) and a C8161 tumor (mean 0.8) scanned in the same day with the same mercury light source. Since this was inactive muscle, we attribute this to State 4 metabolism. The Hb ratio image of the muscle (not shown) was uniform and similar to that of the two tumors.

3.2. DCE MRI

Dynamic contrast enhanced MRI studies were conducted on three aggressive (C8161) and four indolent (A375P) xenografts. Table 1 presents the results of an analysis of these data on the basis of the BOLERO algorithm. Data were analyzed by the FXL-constrained and the FXR-permitted methods described in Methods. The only statistically significant difference in the parameters that were obtained from this analysis was for K_{trans} of the tumor cores obtained by the FXR-permitted method.

Figure 5 compares values of K_{trans} obtained by the FXL-constrained and FXR-permitted methods. This parameter was about twice as great in the indolent tumor as in the aggressive tumor (unpaired student t test $p<0.05$). The FXL-constrained analysis produced no significant difference between any of the parameters of the aggressive or indolent tumors and yielded a mean value of K_{trans} for the indolent tumors that was about half the value obtained by the FXR-permitted method.

For small animal tumor data, the FXR-permitted method yielded better data fits than the FXL-constrained analysis.[10] In addition, in a previous study of the RIF-1 tumor model the FXL-constrained analysis produced a spurious dependence of v_e on the Gd-DTPA concentration (R. Zhou, J.D. Glickson, C.S. Springer, unpublished). Therefore, the FXR-permitted analysis is preferred.

On the basis of the vasculogenic mimicry concept, we would have expected to see a higher K_{trans} in the aggressive than the indolent tumor. However, histological analysis (Figure 4) indicates that there is more blood vasculature in the aggressive tumor, but most of the vasculature appear collapsed and many of them are probably non-functional.

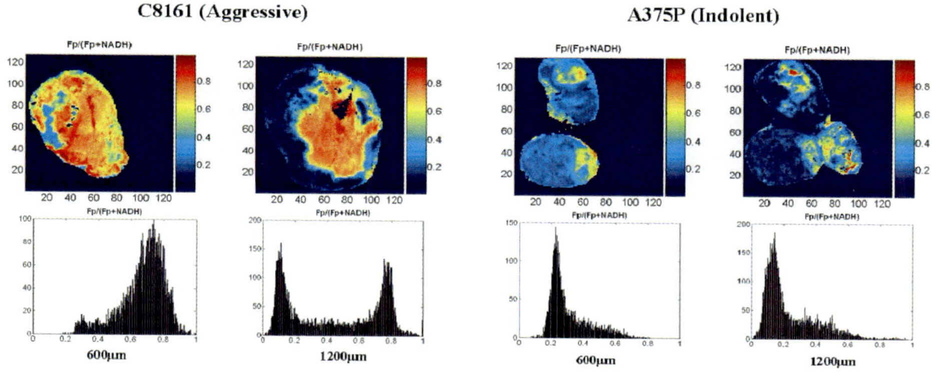

Figure 1. Typical images and histograms of redox ratios of two sequential tissue slices of C8161 versus A375P xenografts at depths of 600 µm (left) and 1200 µm (right).

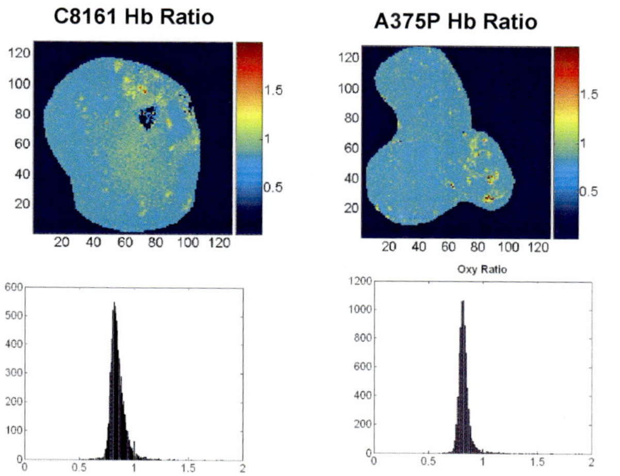

Figure 2. Images and histograms of Hb ratio of the aggressive (left) and indolent (right) melanoma xenografts at a slice depth of 1200 µm from the top surface of the tumors.

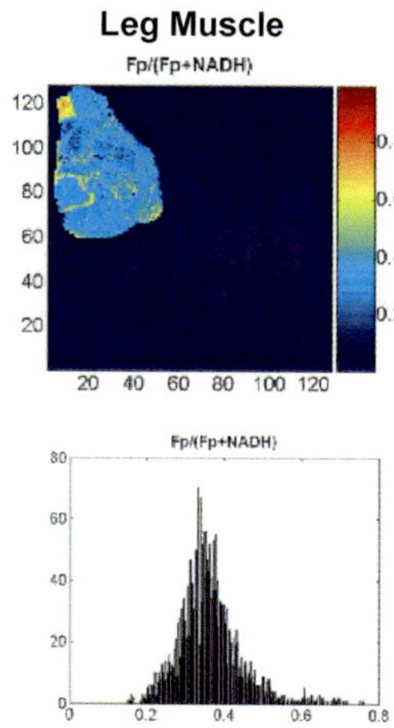

Figure 3. Redox ratios of normal resting mouse muscle at depth of 1600 µm.

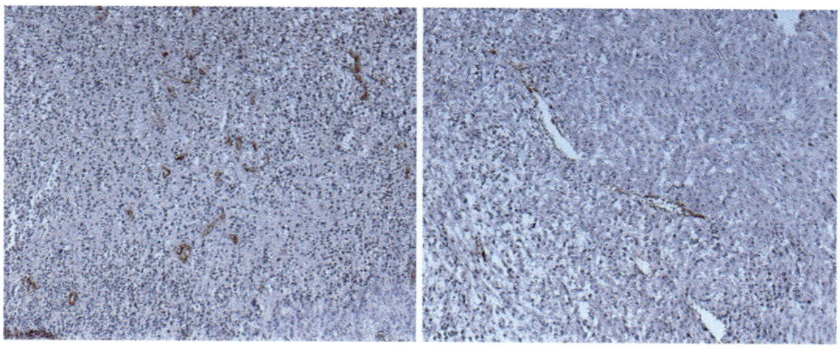

Figure 4. Histological slides of an aggressive C8161 melanoma tumor specimen (left) and indolent melanoma A375P tumor specimen (right). An anti-factor VIII stain has been applied to identify vascular endothelium.

Table 1. Summary of analysis of DCE MRI experiments on indolent (A375P) and aggressive (C8161) human melanoma xenografts. Data were analyzed by the BOLERO algorithm that utilized a fast exchange limit model (i.e., FXL-constrained model) or a fast exchange regime model (i.e., the FXR-permitted model) as described in Methods. Errors represent inter-tumor standard deviations. * Unpaired student t test $p < 0.05$

	Tumor Center		Tumor Rim	
Parameter	C8161 (n=3)	A375P (n=4)	C8161 (n=3)	A375P (n=4)
$K_{trans}(min^{-1})$	0.06±0.03	0.15±0.06*	0.15±0.07	0.30±0.21
v_e	0.14±0.12	0.31±0.08	0.36±0.25	0.36±0.25
$\tau_i(s)$	0.4±0.4	2.1±1.7	1.8±0.2	1.8±0.4

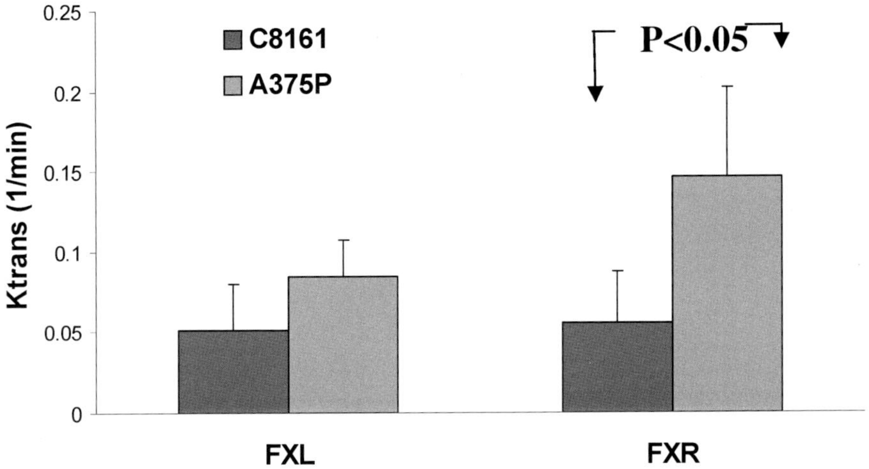

Figure 5. Mean K_{trans} in tumor cores of indolent A375P melanomas (light gray) and aggressive C8161 melanomas (dark) as determined by the FXL-constrained (left) and FXR-permitted (right) methods. Statistical significant ($p<0.05$) difference between these two lines was only detected using the FXR-permitted method.

Therefore, the aggressive tumor may produce more blood vasculature because of the ability of melanoma cells to generate vascular channels, but these blood vascular structures do not appear to lead to an overall improvement in the functional blood exchange of the aggressive tumor. Rather this tumor appears to be more poorly perfused and/or equipped with less leaky vasculature. In addition, the histological analysis revealed extensive regions of gross necrosis in the aggressive but not in the indolent tumor (not shown in Figure 4). This again is consistent with the assignment of the mitochondrial state of the tumor to State 2 (starved) rather than to State 3 (active and well supplied with substrate). However, we recognize that these paradoxical data may be explained by our protracted window of observation and the voluminous number of melanoma cells injected into the mice. Previous experiments, especially with the highly aggressive C8161 melanoma cells, have utilized half the number of tumor cells 1×10^6 for subcutaneous implantation and observed metastasis within 2 weeks.[7] Therefore, the necrosis seen in the current study, along with the associated malfunctional vasculature, may be a consequence of tumor cell overload and an artificially extended window of observation.

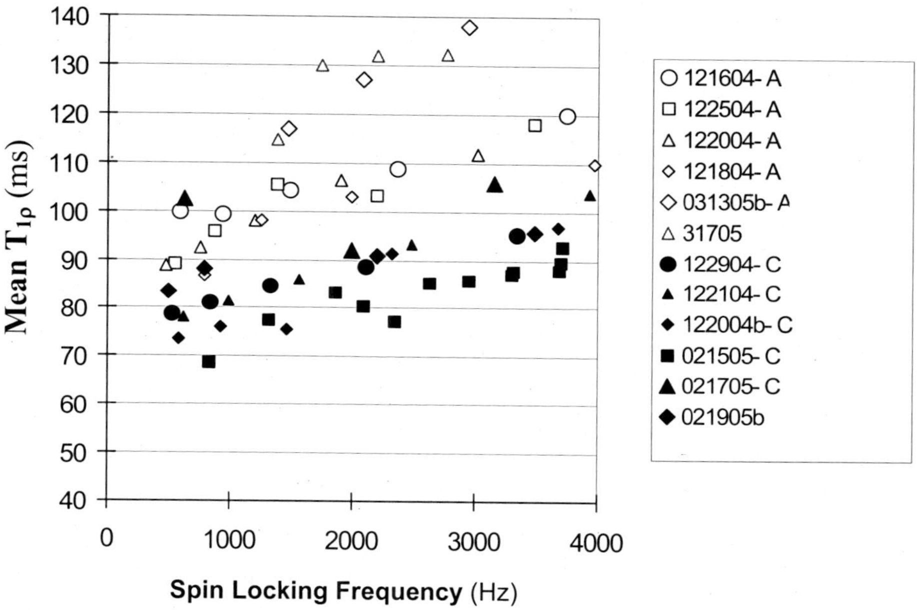

Figure 6. Mean $T_{1\rho}$ values versus the frequency of the spin locking pulse for indolent melanoma A375P (open markers, n=6) and aggressive C8161 melanoma (solid markers, n=6).

Table 2. Comparison of mean $T_{1\rho}$, T_2, T_1 between A375P and C8161 melanoma.

	mean $T_{1\rho}$±STD (ms)			T_2±STD (ms)	T_1±STD (s)
SLF (Hz)	490-1499	1500-2499	2500-3499	-	-
A375P	N=6	n=6	n=4	n=6	n=4
	103±10	113±12	125±12	53±7	2.07±0.03
C8161	N=6	n=6	n=4	n=6	n=4
	83±11	88.5±5	96±8	45±8	2.05±0.03
p (t-test)	<0.01	<0.005	<0.01	>0.05	>0.1

3.3. $T_{1\rho}$ MRI of Melanoma

We have observed significant differences between the average $T_{1\rho}$ values of the indolent A375P and aggressive C8161 tumors at SLFs ranging from 490-4000Hz. Figure 6 displays a plot of mean $T_{1\rho}$ values versus spin locking frequency for twelve melanomas, six indolent A375P tumors and six aggressive C8161 tumors. When the SLF was above 800 Hz, $T_{1\rho}$ tended to increase with SLF, and the indolent tumor exhibited distinctly longer relaxation times than the aggressive tumor. When the SLF was below 800 Hz, the separation was still quite distinct, though a couple of data points overlapped.

To determine if the differences between A375P and C8161 $T_{1\rho}$ values were statistically significant, the $T_{1\rho}$ data were divided into three subgroups according to their SLF values: 490-1499, 1500-2499, 2500-4000Hz. For each subgroup, the mean value and standard deviation of $T_{1\rho}$ were calculated for the A375P and C8161 tumors and are presented in Table 2. T-test analysis indicates that the differences were statistically significant with $p<0.01$ or lower (Note: although the data show that $T_{1\rho}$ had a moderate dependence on SLF, the trend of dependence was approximately linear and similar between the aggressive and indolent tumors. The SLF-dependent $T_{1\rho}$ variation was also significantly smaller than the inter-tumor variation. Therefore, on first order approximation, it was justified to perform the statistical comparison of the subinterval mean $T_{1\rho}$ difference between the two types of tumors, and the moderate dependence of $T_{1\rho}$ on SLF should not have caused significant errors in the t-test analysis). The average T_2 values, 50±7 ms for the A375P tumors and 43±10 ms for C8161 tumors, were not significantly different on the basis of unpaired t-test analysis. Neither were the differences in tumor mean T_1 values significant ($p>0.1$) (Table 2). Therefore only the $T_{1\rho}$ relaxation times were suitable for distinguishing between these aggressive and indolent tumors.

Increases in $1/T_{1\rho}$ have been observed to correlate with the macromolecular concentration,[17,18] which might be associated with the substantial necrosis noted in the aggressive tumor, although a correlation of decreased $T_{1\rho}$ with tissue necrosis is not well established in the literature. Another possibility is that the indolent tumor was more acidic than the aggressive tumor (due to more glycolysis in the indolent tumor). A decrease in proton exchange rate that would be expected in a more acidic environment could lead to an increase in $T_{1\rho}$.

4. CONCLUSIONS

Three methods potentially applicable to human patients have been demonstrated to distinguish between aggressively metastatic and indolent human melanoma xenografts. One of these methods (optical imaging) can be applied to frozen biopsy specimens to obtain 3D high resolution images of tumor heterogeneities, whereas the other two can be utilized noninvasively in humans. The optical method also showed a marked increase in the Fp redox ratio in the core of the aggressive tumor. In some slices the rim of the aggressive tumor seems to have had a much lower redox ratio (higher NADH level). Since the DCE MRI data indicate lower functional blood exchange (perfusion/permeability) in the core of the aggressive tumor, the low Fp redox ratio was interpreted as indicative of State 2 mitochondrial metabolism – i.e., the core appears to have been depleted of substrate. A more poorly perfused core of the C8161 tumor would be expected to secrete VEGF and other cytokines that could facilitate metastasis.

Note that intertumor heterogeneity was quite high, as particularly indicated by the observations of DCE and optical imaging; thus, more studies are needed to confirm these preliminary observations. The effect of tumor mass and overall tumor body burden on the mitochondrial state, the perfusion characteristics and $T_{1\rho}$ relaxivity of these tumors needs to be examined in order to permit a meaningful comparison with human cancer that normally is associated with lower tumor body burden and, therefore, better perfusion than the murine models.

Three other melanoma cell lines with intermediate levels of metastatic potential in the human host have been obtained from Dr. Hendrix's laboratory and will be examined as xenografts. Future work will be directed to comparison of xenografts of these sublines with those that have already been characterized. Additionally, extension to other forms of human cancer is contemplated. If the difference between an indolent and an aggressive melanoma observed in this study can be generalized to other cancer cell lines and correlated with the graded levels of tumor metastatic potential, these imaging techniques would provide valuable methods for cancer diagnosis. The availability of reliable methods for predicting the aggressiveness of human cancers would have enormous impact on the clinical management of this disease.

5. ACKNOWLEDGEMENTS

This research was supported by NIH grant P01-CA56690-09A2 (D.B. Leeper, PI, J.D. Glickson, project director), P50-CA 093372 (M. Herlyn, PI), UO1-CA105490 (L.A. Chodosh, PI). The $T_{1\rho}$-weighted MRI was assisted by Dr. Ravinder Regatte and Dr. Ravinder Reddy. Mr. David Nelson in preparation of the manuscript is acknowledged.

6. REFERENCES

1. Folberg R, Hendrix MJ & Maniotis AJ. Vasculogenic mimicry and tumor angiogenesis. *American Journal of Pathology* **156**(2), 361-381 (2000).
2. Hendrix MJ, Seftor EA, Hess AR & Seftor RE. Vasculogenic mimicry and tumour-cell plasticity: lessons from melanoma. *Nature Reviews Cancer* **3**(6), 411-421 (2003).
3. Maniotis AJ, Folberg R, Hess A, Seftor EA, Gardner LM, Pe'er J, Trent JM, Meltzer PS, Hendrix MJ. Vascular channel formation by human melanoma cells in vivo and in vitro: vasculogenic mimicry. *American Journal of Pathology* **155**(3), 739-752 (1999).
4. Maniotis AJ, Chen X, GArcia C, DeChristopher PS, Wu D, Pe'er J, Holberg R. Control of melanoma morphogenesis, endothelial survival, and perfusion by extracellular matrix. *Laboratory Investigation* **82**(8), 1031-43 (2002).
5. Ruf W, Seftor EA, Petrovan RJ, Weiss RM, Gruman LM, Marganyan NV, Seftor EA, Nagle RB. Differential role of tissue factor pathway inhibitors 1 and 2 in melanoma vasculogenic mimicry. *Cancer Research* **63**(17), 5381-9 (2003).
6. Hendrix MJ, Seftor EA, Chu YW, Seftor RE, Magle RB. Coexpression of vimentin and keratins by human melanoma tumor cells: correlation with invasive and metastatic potential. *Journal of the National Cancer Institute* **84**(3), 165-174 (1992).
7. Welch DR, Bizi JE, Miller BE, Conoway D, Seftor EA, Yohan KH, Gilmore LB, Seftor RE, Nakajima M, Hendrix MJ. Characterization of a highly invasive and spontaneously metastatic human malignant melanoma cell line. *International Journal of Cancer* **47**(2), 227-237 (1991).
8. Quistorff B, Haselgrove JC & Chance B. High resolution readout of 3-D metabolic organ structure: An automated, low-temperature redox ratio-scanning instrument. *Anal. Biochem.* **148**389-400 (1985).
9. Kozlowski JM, Hart IR, Fidler IJ & Hanna N. A human melanoma line heterogeneous with respect to metastatic capacity in athymic nude mice. *Journal of the National Cancer Institute* **72**(4), 913-7 (1984).

10. Zhou R, Pickup S, Yankeelov TE, Springer CS & Glickson JD. Simultaneous measurement of arterial input function and tumor pharmacokinetics in mice by dynamic contrast enhanced imaging: effects of transcytolemmal water exchange. *Magn. Reson. Med.* **52**248-57 (2004).
11. Yankeelov TE, Rooney WD, Li X & Springer CS Jr. Variation of the relaxographic "shutter-speed" for transcytolemmal water exchange affects the CR bolus-tracking curve shape. *Magnetic Resonance in Medicine* **50**(6), 1151-69 (2003).
12. Wheaton AJ, Borthakur A, Kneeland JB, Regatta RR, Akella SV, Reddy R. In vivo quantification of $T_{1\rho}$ using a multislice spin-lock pulse sequence. *Magnetic Resonance in Medicine* **52**(6), 1453-1458 (2004).
13. Borthakur A, Wheaton AJ, Gongontas AJ, Akella SV, Regatta RR, Charagundla SR, Reddy R. In vivo measurement of $T_{1\rho}$ dispersion in the human brain at 1.5 tesla. *Journal of Magnetic Resonance Imaging* **19**(4), 403-409 (2004).
14. Chance B, Schoener B, Oshino R, Itshak F & Nakase Y. Oxidation-reduction ratio studies of mitochondria in freeze-trapped samples. NADH and flavoprotein fluorescence signals. *Journal of Biological Chemistry* **254**(11), 4764-4771 (1979).
15. Chance B. in *Flavins and Flavoproteins* (ed. Slater, E. C.) 498-510 (Elsevier, Amsterdam, 1966).
16. Chance B & Baltscheffsky H. Respiratory Enzymes in Oxidative Phosphorylation. *J. Biol. Chem.* **233**(2), 736-739 (1958).
17. Virta A, Komu M & Kormano M. T1rho of protein solutions at very low fields: dependence on molecular weight, concentration, and structure. *Magnetic Resonance in Medicine* **37**(1), 53-57 (1997).
18. Duvvuri U, Golberg AD, Krantz JK, Hoang L, Reddy R, Wehrli FW, Wand AJ, Englander SW, Leigh JS. Water magnetic relaxation dispersion in biological systems: the contribution of proton exchange and implications for the noninvasive detection of cartilage degradation. *Proceedings of the National Academy of Sciences of the United States of America* **98**(22), 12479-12484 (2001).

ANALYSIS OF SDHD AND MMP12 IN AN AFFECTED SOLAR KERATOSIS AND CONTROL COHORT

N.A. Lintell*, D.J. Maguire*, L.R. Griffiths#, and M. McCabe*

Abstract: The incidence of Squamous Cell Carcinoma (SCG) is growing in certain populations to the extent that it is now the most common skin lesion in young men and women in high ultraviolet exposure regions such as Queensland. In terms of incidence up to 40% of the Australian population over 40 years of age is thought to possess the precancerous Solar Keratosis (SK) lesion and with a small, but significant, chance of progression into SCC, understanding the genetic events that play a role in this process is essential.

The major aims of this study were to analyse whole blood derived samples for DNA aberrations in genes associated with tumour development and cellular maintenance, with the ultimate aim of identifying genes associated with non-melanoma skin cancer development. More specifically the first aim of this project was to analyse the SDHD and MMP12 genes via Dual-Labelled Probe Real-Time PCR for copy number aberrations in an affected Solar Keratosis and control cohort.

It was found that 12 samples had identifiable copy-number aberrations in either the SDHD or MMP12 gene (this means that a genetic section of either of these two genes is aberrantly amplified or deleted), with five of the samples exhibiting aberrations in both genes. The significance of this study is the contribution to the knowledge of the genetic pathways that are malformed in the progression and development of the pre-cancerous skin lesion Solar Keratosis.

1. INTRODUCTION

Squamous Cell Carcinoma is now the most common skin lesion in young men and women in high ultraviolet exposure regions such as Queensland[1]. In terms of incidence, up to 45% of the Australian population over 40 years of age is thought to possess the

* School of Biomolecular and Biomedical Science and #School of Medical Science, Griffith University, Queensland 4111, Australia.

precancerous Solar Keratosis lesion and with a small but significant chance of progression into SCC, understanding the genetic events that play a role in this process is essential[2]. Apoptosis is a cellular process that removes diseased and damaged cells and it is the malfunction of this pathway that has been fundamentally linked to the initiation of carcinogenesis[3].

Although there have been a number of studies investigating potential genetic risk factors for precancerous conditions, to date there have been no published studies focussing on genes involved in the energy production and cell maintenance pathways. Mitochondria are responsible for the initiation and regulation of apoptosis in addition to being the prime energy progenerator[4]. It was observed by Warburg that some cell types decrease oxidative phosphorylation and increase glycolytic energy production during tumourigenesis[5]. Warburg *et al.* also first reported physiological changes of mitochondria in cancer cells [6]

Mutations in complex II of the electron transport chain (ETC) have a strong association with inherited paragangliomas[7, 8]. In 10–15% of the cases, these usually benign neuroectodermal tumours are inherited in an autosomal dominant fashion with incomplete penetrance. The mutations in SDHB, SDHC, and SDHD appear to be responsible for the majority of familial paragangliomas[7, 8] and also for a significant fraction of non-familial tumours, including phaeochromocytomas (tumours of the adrenal medulla)[7, 8]. The SDHD protein is thought to be part of the oxygen-sensing system in paraganglionic tissue and its loss may lead to chronic hypoxic stimulation and cellular proliferation[9].

In this role, SDHD acts as a tumour suppressor gene in paraganglionic tissue and has also been linked to the stimulation of the angiogenic pathway in these types of neoplasm's [9].

Two members of the matrix metalloproteinase family (MMP), which perform the function of digesting all matrix macromolecules such as collagen and elastin, have been linked to the remodelling of elastotic areas in sun-damaged skin[10]. The two MMP molecules, MMP-7and and MMP-12, have found to be over-expressed in aggressive SCCs and granulomatous skin disorders, respectively[11, 12], with MMP-12 also theorised to be associated with two types of inflammatory skin disorders[12].

2. METHODOLOGY

Table 1 shows the primer sequences devised for analysing specific exons of the genes in question, along with their associated PCR conditions. These gene sets were analysed for copy-number aberrations via Dual-Labelled Real-Time PCR. The Dual-Labelled Probe Real-Time PCR technique is a relatively inexpensive, but reliable way to scan large sample populations for putative gains and losses in a genetic area of interest. Samples are assumed to be aberrant if they differ by two or more threshold cycles (Ct) outside the putative wild-type range as per the standard literature[13,14,] with samples identified as anomalous portrayed in Tables 2 and 3.

Table 1. Primer sequences, $MgCl_2$ concentrations, and melt temperatures.

Gene	Oligo. type	Oligonucleotide Sequence	Conditions
SDHD-DL	Forward	GGTCAGACCTGCTCATATCTC	2.5uM, 64°C
SDHD-DL	Reverse	TGGCTCGGTGACAAGTGTATG	2.5uM, 64°C
MMP12-DL	Forward	GGTCTGAAAGTGACCGGGCA	2.5uM, 64°C
MMP12-DL	Reverse	ATGGACATCGGGGACTCCAC	2.5uM, 64°C
Tubulin-α-8	Forward	GCAGCCAACAACTATGCCCG	2.5uM, 64°C
Tubulin-α-8	Reverse	CTTCCGTATGCGGTCCAGCA	2.5uM, 64°C
Tubulin-α-8	Probe	6-Fam-TCAATGCTCTCCTTGCCCACCGTGTAGT-BHQ1	68°C
SDHD	Probe	Joe-TGCTGCACTCCACACCATTTGGGATAG-BHQ1	68°C
MMP12	Probe	Rox-GAGGTGCGTGCATCATCTCCAGGGTAG-BHQ2	68°C

A standard magnesium titration was performed on the three sets of primers outlined in Table 1 *via* the Hybaid OminGene PCR machine with a mineral oil overlay. For each of the primer sets for the dual-labelled probes, a standard reaction master mix was made. This reaction master mix contained: 1X BioTaq reaction buffer, 0.8 μM dNTPs, 0.5μM primer and 1.0 unit of BioTaq polymerase. The master mix included a set volume of 2.5μM $MgCl_2$ following the design specificities of the oligonucleotides that were used in the experiment. The cycle conditions are as follows: 95°C 3 Min., 45x (95°C, 20 sec., 64°C, 20 sec., 68°C, 20 sec., 72°C, 30 sec.)

3. RESULTS

Table 2 shows the aberrant sample number, the gender of the subject and the SK status, and the type of aberration (either gain or loss). The examples that possess bold descriptions in the aberrant gene columns indicate the presence of a sample exhibiting copy-number aberrations in both the SDHD and MMP12 genes.

There were 12 samples identified as aberrant for either exon two of the MMP-12 gene or exon three of the SDHD gene, with five of these samples exhibiting anomalies in both the relevant genetic loci. To ensure that non-specific products were not influencing (their influence on the reaction is moderate, but still needs to be taken into account) the data derived from the Dual-Labelled Probe Real-Time PCR runs, agarose gel electrophoresis was performed on each of the 12 samples. From these gels (data not shown), it was evident that there were no unspecific products formed during the reaction cycles.

Table 2. Summary of SK DNA Dual-Labelled probe aberrants.

Sample	SK Status and gender	SDHD Aberration	MMP12 Aberration
47	Affected male	Genetic loss	–
104	Control male	**Genetic gain**	**Genetic gain**
207	Control male	–	Genetic loss
245	Control female	**Genetic loss**	**Genetic loss**
296	Control female	–	Genetic gain
310	Affected male	–	Genetic gain
316	Affected male	**Genetic gain**	**Genetic gain**
357	Affected female	–	Genetic gain
397	Control male	Genetic loss	–
408	Affected female	–	Genetic gain
455	Control male	**Genetic gain**	**Genetic gain**
461	Control male	**Genetic loss**	**Genetic loss**

Table 3 shows the threshold cycle (Ct) variances between run (inter-run) and within-run (intra-run). The parameter that has the greatest influence on Dual-Labelled Real-Time PCR is the intra-run variance, with none of the intra-run Ct variances above half a cycle, with the majority existing around 0.1 of a cycle variance. This indicates this analytical technique is quite precise and has high reproducibility for the identification of copy-number aberrations.

The Dual-Labelled Probe Real-Time PCR method of diagnosis proved adequate in terms of reproducibility and cost, with the average cost of a probe and primer set being AUD $ 450.00. Each probe and primer set gives a minimum of 800 uses, with all the probes utilised in this study having considerably more repetitions left in its sample volume. Tubulin-α-8 as the housekeeper showed extremely high reproducibility in terms of the Real-Time PCR quantitation curves it produced, which were clear and easily discernible.

Table 3. SK DNA Real-Time PCR inter-run and intra-run Ct variances.

Sample	Tubulin intra-run	Tubulin inter-run	SDHD intra-run	SDHD inter-run	MMP12 intra-run	MMP12 inter-run
47	0.173	0.889	0.141	0.141	–	–
104	0.255	0.255	0.071	0.283	0.131	0.434
207	0.141	0.141	–	–	0.141	0.141
245	0.176	0.161	0.169	0.402	0.424	0.424
296	0.244	0.439	–	–	0.375	0.759
310	0.050	1.19	–	–	0.160	0.460
316	0.040	1.26	0.209	0.183	0.127	0.127
357	0.131	1.03	–	–	0.020	0.863
397	0.134	0.134	0.180	0.284	–	–
408	0.138	0.243	–	–	0.240	0.240
455	0.173	0.834	0.219	1.30	0.070	0.571
461	0.010	0.010	0.290	0.290	0.064	0.064

4. DISCUSSION

Lesions of the skin originating from non-melanotic cells occur at a rate significantly higher than other neo-plastic malignancies within the Australian population[15]. However, the knowledge of their genomic and proteomic manifestations falls below that of breast and colon cancer[15]. These types of tumours, non-melanotic skin cancers (NMSC), are ideal for clinical research because of the abundance of medical samples available for study and their accessibility on the human body. One type of NMSC, SCC, has a dermatologically identifiable pre-cancerous state, in which it can proliferate into a malignant lesion or regress into a benign tissue state[2]. This stage of this disease is known as Solar Keratosis (SK), but as yet no clear genetic shifts have been identified in this stage of the disease for a progression into SCC, other than mutations in the gene encoding for the p53 cell cycle mechanism[2].

This project analysed chromosomal regions that contained both a nuclear-encoded electron transport chain gene and an oncogene/tumour suppressor gene associated with non-melanoma skin cancer for relatively large-scale mutations. In order to do this, a real-time PCR approach was optimised, encompassing a quantitative curve component with threshold cycle comparison[13, 14].

This study was focussed on the 11q23 locus, a chromosomal region that has previously been found to contain regional gains and losses in a comparative genomic hybridisation (CGH) study involving samples extracted from both SK and SCC lesions[15]. The importance of this particular finding is that the SK lesion has long been identified as part of a continuum with SCC, both histopathologically and dermatologically[2]. What is even more important is that the regional losses were restricted to the 11q22-qter region, whilst the regional gain was found to exist on the entire 11q arm. This means that the gains identified in this study would not be part of the 11q13 amplicon spectrum; a region identified in numerous CGH studies involving cancerous tissue, such as breast cancer, as being aberrant[15].

There were 12 samples with an anomalous nature for either of the SDHD and MMP-12 genes out of a total population of 279, with five of the samples exhibiting aberrations in both the SDHD and MMP-12 genes. The SDHD component possessed two samples that exhibited genetic losses, with no samples showing a gain at exon three for this gene. Exon two of the MMP-12 gene exhibited four samples with a gain characteristic and one sample showing a genetic loss. The five samples possessing copy-number abnormalities in both genes showed a gain occurring in three samples and a genetic loss in two. These genes have been associated with tumour development as well as inflammation. However, they are of the suppressor type, in which both copies of the gene need to be malformed. Thus, their presence in relative equal numbers in the control and affected population (seven and five, respectively) is not an unexpected result. This study, therefore did not find any indication that these two genes play a significant role in the development of the precancerous SK lesion, particularly in terms of copy-number aberrations. These types of aberrations are a key component in the inactivation of tumour suppressor genes, a fundamental process vital to tumour initiation and promotion[16].

A future research interest is investigating the role SDHD plays in the oxygen-sensing system in paraganglionic tissue and how its loss may lead to chronic hypoxic stimulation and cellular proliferation. Whether this is the only type of carcinoma where SDHD plays such a role or not would be a key question for research into neoplasms such as SK and SCC. Another would be whether SDHD is the only such ETC subunit protein that acts as an oxygen sensor in hypoxic situations, such as inflammation of elastotic areas in sun-damaged skin, an aspect of macromolecular regulation that MMP-12 plays a central role in or not.

The cost of sequencing numerous genetic areas in a large sample cohort is usually lacking in feasibility in small studies such as this one. Another precise method of analysing specific genes can be screening the samples first using a non-specific fluorophore, with both the quantitation and melt-curve methods. Analysing the putative aberrant population from that study via single-labelled and dual-labelled probes would then give the project a more definitive atypical cohort, in terms of identifying DNA mutations and polymorphisms. The first method can accurately analyse a sample for amplicon melting anomalies, which can indicate the presence of a nucleotide aberrations down to the single-nucleotide level. The second method is a precise way to detect the presence of gross copy-number aberrations in the genetic area of interest[13, 14].

This project was the first, as far as a three-year literature review across numerous scientific databases could ascertain, to analyse precancerous lesions for abnormalities in genes that encode for subunits of the ETC. Another point of interest is that the DNA utilised in this project did not come from an epithelial source, rather it was extracted from whole blood samples taken from a control and affected SK cohort. Therefore, the genetic malformations detected in this project have purportedly wider implications in terms of systematic affectations contributing to tumour initiation and promotion.

The Dual-Labelled Probe Real-Time PCR method proved to be a precise technique in the detection of copy-number aberrations as shown in Table 3. A prior screening of putative samples via a non-specific fluorophore such as LCGreen I minimises unproductive wasting of probe samples. Searches relating CGH with the type of disease also gives insight into chromosomal areas that might be associated with disease development and progression. If CGH results are combined with Dual-Labelled Probe Real-Time PCR results in terms of genetic gains and losses, then a large and small-scale chromosomal locus picture of a disease model can be theorised. This is particularly feasible in light of the large database of allelic variants located on the OMIM website with regards to certain genes and their association with the development of specific diseases. Apart from studies previously published by our lab[17, 8], there have been no published reports centred on Electron Transport Chain subunit gene abnormalities in samples taken from individuals affected by Solar Keratosis. The identification of aberrations in the ETC genes analysed in DNA extracted from lymphocyte cells is therefore an unexpected result and gives rise to future directions with regard to the development of this neo-plastic skin lesion.

5. REFERENCES

1. Frost C, Williams G and Green A. High incidence and regression rates of solar keratoses in a Queensland community. *J. Invest. Dermatol.* **115** (2000), pp. 273-277.
2. Rosen RH and Studniberg H. Solar keratoses: analysis in a dermatological practice in Australia *Nutrition.* **44** (2003), pp: 34-39.
3. Gosslau A and Chen KY. Nutraceuticals, apoptosis, and disease prevention. *Oral Oncology.* **20** (2004), pp. 95-102.
4. Modica-Napolitano JS and Singh KK. Mitochondrial dysfunction in cancer. *Mitochondrion.* (2004), pp. 1-8.
5. Warburg O. The origin of cancer cells. *Science.* **123** (1956), pp. 309-314.

6. Warburg O, Posenor K and Nehelein E. über den Stoffwechsel der Carcinomzelle. *Biochem Z.* **152** (1924), pp. 309-345.
7. Zeviani M, Spinazzola A and Carelli V. Nuclear genes in mitochondrial disorders. *Curr. opin. genetics dev.* **13** (2003), pp. 262-270.
8. Zeviani, M and Di Donato S. Mitochondrial disorders. *Guarantors of Brain.* **127** (2004), pp. 2153-2172.
9. Gimenez-Roqueplo AP, Favier J, Rustin P, Mourad JJ, Plouin PF, Corvol P, Rotig A and Jeunemaitre X. The R22X mutation of the SDHD gene in hereditary paragangiloma abolishes the enzymatic activity of complex II in the mitochondrial respiratory chain and activates the hypoxia pathway. *Am. J. Hum. Genet.* **69** (2001), pp. 1186-1197.
10. Saarialho-Kere U, Kerkla E, Jeskanen L, Hasan T, Pierce R, Starcher B, Raudasoja R, Ranki A, Oikarinen A and Vaalamo M. Accumulation of matrilysin (MMP-7) and macrophage metalloelastase (MMP-12) in actinic damage. *J. Invest. Dermatol.* **113** (1999), pp. 664-672.
11. Karelina TV, Goldberg GI and Eisen AZ. Matrilysin (PUMP) correlates with dermal invasion during appendageal development and cutaneous neoplasia. *J. Invest. Dermatol.* **103** (1994), pp. 482-487.
12. Vaalamo M, Kariniemi A-L, Shapiro SD and Saarialho-Kere U. Enhanced expression of human metalloelastase (MMP-12) in cutaneous granulomas and macrophage migration. *J. Invest. Dermatol.* **112** (1999), pp. 499-505.
13. Bernard PS and Wittwer CT. Real-Time PCR technology for cancer diagnostics. *Clin. Chem.* **48** (2002), pp. 1178-1185.
14. Mocellin S, Rossi CR and Marincola FM. Quantitative Real-Time PCR in cancer research. *Arch. Immunol. ther. Ex.* **51** (2003), pp. 301-313.
15. Ashton KJ, Weinstein SR, Maguire DJ, and Griffiths LR. Molecular cytogenetic analysis of basal cell carcinoma DNA using comparative genomic hybridisation. *J of Invest. Dermatol.* **117**, No. 3 (2001), pp. 683-686.
16. Bertram JS. The molecular biology of cancer. *Mol. Aspects Med.* **21** (2001), pp. 167-223.
17. Maguire, D.J., Lintell, N.A., McCabe, M., Griffiths, L.R. & Ashton, K.S. Genomic and phenomic correlations in the respiration of non- melanotic skin cancers. *Advances in Experimental Medicine and Biology.* **540** (2003), pp. 251-256.
18. Lintell N. A., Maguire, D. J., McCabe, M. & Griffiths, L.R. Focussing on genomic and phenomic correlations in the respiration of non-melanotic skin cancers. *Advances in Experimental Medicine and Biology.* **566** (2004), pp. 375-380.

WYMAN'S EQUATION AND OXYGEN FLUX THROUGH THE RED CELL

Michael McCabe* and David J. Maguire*

Abstract: Wyman's equation of 1966 [1] describes the facilitation of flux of a reversibly bound substrate such as oxygen, consequent on the translational diffusion of the binding protein (the carrier). While Wyman's equation [1], or some modification of it such as that by Murray [2], may provide a realistic description of the flux of oxygen through a dilute solution of haemoglobin (see also Wittenburg [3, 4]), it is unlikely to be the complete explanation, nor even the basis, for oxygen transport through the intact red cell. The mature erythrocyte contains approximately 350g/l haemoglobin, and while this suggests that only 35% of the available water volume is actually occupied by the protein, the remaining 65% is unavailable for protein translational diffusion due to the mutual exclusion of the haemoglobin molecules. For this reason we have examined other possible mechanisms whereby haemoglobin may facilitate the translational diffusion of oxygen within the erythrocyte. Possible alternatives include rotational diffusion by the haemoglobins, intracellular shuffling of haemoglobins due to shape changes by the erythrocyte, and haemoglobin rotations and oxygen exchange consequent on the charge change which accompanies substration and desubstration of the haemoglobin molecule. Finally the dipole interactions are shown to generate significant intermolecular attractions between adjacent haemoglobins.

1. INTRODUCTION

Macromolecular crowding is ubiquitous inside cells and is predicted by well established theory to have large consequences on translational diffusion. In recent years several significant studies have demonstrated that translational diffusion of globular proteins inside cells is extremely restricted (see for example the work of Banks and Fradin [5]; Martine Arrio-Dupont et al. [6]; Jacobson and Wojcieszyn [7]). Despite the well known protein immobilising effects of crowding within cells, the idea still persists in the

* Genomics Research Centre, Griffith University, Nathan, Queensland, Australia

literature that translational diffusion of proteins is a useful and exploited function, (see for example Wittenburg [3] and Wittenburg and Wittenburg [4, 8]). Because intracellular translational diffusion of globular proteins is normally extremely limited, we suspect that many globular proteins have evolved to function effectively within the crowded milieu of the cell interior by adapting their functions and perhaps even aspects of their shape so as to take advantage of the free rotational movements allowed by spheroidal globular proteins, rather than try to take advantage of translational movements, which are essentially very restricted or denied them.

The exclusion volume of proteins is due to two phenomena; firstly the rotation of non spherical proteins constantly sweeps out a space somewhat larger than the true protein volume, thus there is an effective radius of the rotating particle, which is half the longest diameter. Secondly, the centres of adjacent identical protein molecules may not approach each other closer than a distance of this longest diameter. We show here that translational diffusion is scarcely possible within the crowded space of the red cell, partly because of the severe protein crowding, and partly as a consequence of surprisingly strong intermolecular dipole attractions which act to stabilise a quasi-lattice in solution. Rotational diffusion is largely unimpaired, although some orientations are preferred within the lattice like structure.

2. SPACE AVAILABLE FOR HAEMOGLOBIN INSIDE THE ERYTHROCYTE

The volume of each haemoglobin molecule in solution is given by VHb = vM/N where v is the partial specific volume (approximately equivalent to the reciprocal of the density of the solute). M is the molecular weight, and N the Avogadro number. Assuming that v for Hb is 0.75 then

$$\text{VHb} = 8.01 \times 10^{-20} \text{ ml.}$$

Alternatively, the volume per Hb molecule can be approached using its known measured dimensions. Haemoglobin is believed to be an ellipsoid with an axial ratio of 1.18 and diameters of 55A and 65A (Bragg et al,1952 [9]; Ingram et al,1956 [10]). Making the simplifying assumption that Hb is spherical, and from the equation

$$v = \frac{4}{3} r^3$$

where r = 27.5A, then $v = 8.7 \times 10^{-20}$ ml

However we should take the larger axis if we believe that the haemoglobin molecule may tumble around any axis. In this case the effective volume swept out by the haemoglobin will be equivalent to a sphere of radius 32.5A. This gives a volume of almost 15×10^{-20} ml.

The solution space available per haemoglobin molecule inside a solution containing 0.35g Hb/ml can be readily calculated. Available volume per Hb molecule is given by 1/n where n is the number of Hb's per ml. For a solution containing 0.35g/g of Hb, there will be N/M x 0.35 molecules, where N is the Avogadro number and M the molecular weight. The above calculation shows that each Hb molecule has available to it a volume of 30×10^{-20} ml.

Thus a half of the total red cell interior volume is taken up by the cavities swept out by the tumbling (rotational Brownian movements) of the haemoglobins. The remaining half is distributed among the interstices of these tumbling spheres, and it can be shown that the minimum surface to surface distances between these spheroids of rotation is approximately 5 Angstrom.

Taking 8×10^{-20} ml as an approximate minimum value for the Hb molecular volume, we can calculate the minimum excluded volume as 8 times this, ie 64×10^{-20} ml.

This result shows that each Hb molecule within the red cell can only occupy a volume approximately half of the excluded volume. This suggests that the haemoglobin molecules each occupy a limited and fixed space within a three-dimensional lattice-like structure where translational diffusion is essentially forbidden. Thus although Wyman's equation may provide a reasonable fit for the experimental data of oxygen flux through a dilute solution of haemoglobin or myoglobin, it cannot offer a solution to the problem of facilitated diffusion of oxygen in such concentrated protein solutions as are found within the red cell.

3. ROTATIONAL DIFFUSION OF THE HAEMOGLOBIN AS A TRANSLATING MECHANISM FOR THE FLUX OF OXYGEN

Preston et al. [11] have shown that the rotational diffusion of globular and essentially spherical macromolecules, such as haemoglobin, is virtually unimpeded at concentrations of macromolecules which effectively prevent translational diffusion or sedimentation. It seems that the rotational motions consequent on Brownian motion constantly sweep out the cavity in which the protein sits, permitting only water or other similarly small molecules to remain in occupation. Consequently, rotational motions of proteins in concentrated solutions occur with almost the same freedom as in dilute solutions.

We have demonstrated previously [12] that almost all possible rotations of a carrier carrying a reversibly binding substrate, will have the consequence of translating the substrate down its concentration gradient. This is because the great majority of the temporary attachments between carrier and substrate are formed on the high concentration side of the carrier, thus almost any subsequent rotation has the consequence of moving the substrate down its gradient.

Thus while translation of the haemoglobin carrier may not proceed, the translation of the carried oxygen could proceed via the rotation of the carriers. We conceive the flux of oxygen from say the centre of an oxygenated erythrocyte as proceeding by a series of discrete jumps of the diffusant, from haem sites, either immediately onto an adjacent vacant haem, or by way of available spaces in the surrounding water.

Even when the red cell has just completed a transit through the normal healthy functional lung, it is 97-98% saturated. Thus, on average, 12% of the "oxygenated" haemoglobin molecules carry a vacant binding site for oxygen. If these empty binding sites are randomly distributed among the haemoglobin molecules, then 1 in every 9 or 10 haemoglobin molecules will be carrying a vacant binding site for oxygen. If spheres are close packed in a three-dimensional lattice, then each sphere is immediately adjacent to 12 spheres in the lattice. Thus each haemoglobin molecule with 4 oxygenated binding sites is immediately adjacent to at least one other haemoglobin molecule which has a vacant haem binding site. If the haemoglobinmolecules are undergoing rotational diffusive jumps, then a bound oxygen is often adjacent to a vacant haem site on an adjacent haemoglobin molecule, when a diffusive jump for the oxygen, from one haemoglobin molecule to the next, becomes most feasible.

4. THE DIPOLES OF HAEMOGLOBIN

Most proteins carry significant dipoles, since they are a measure of the asymmetry of charge distribution throughout the molecule. For a protein such as Hb the effective dipole will be the sum of a number of dipoles as follows; the permanent dipole (which will not change as long as the pH remains constant), an induced dipole (consequent of its position adjacent to other dipoles or due to its position in electro-magnetic fields), a fluctuating dipole (as the electron clouds migrate around the protein), and a dipole which may change consequent on an act of substration or desubstration. In general, the closer the proteins are together, then the more significant will become the effective dipole-dipole bonds. For two large spheroidal particles close together (ie when the distance separating the two surfaces is less than the radius of the spheroids) then the dipole attractive forces fall off as the inverse of the *first power* of the separation distance. This relative independence of distance is due to the geometry consequent on large particle size, which is in contrast to small molecules or atoms, where the force of interaction falls off very rapidly with increasing distance according to an inverse sixth power law (see Setlow and Pollard [13], for a fuller discussion). For molecules or particles substantially larger than Hb, then the interatomic attractions, which are the basis for the dipole forces, are no longer in phase, due to the increasingly significant time lag of the fluctuating dipoles, and the interaction energy becomes small. Thus haemoglobin falls within the range of particle size for optimum dipole interaction. Additionally, for Hb inside the red cell, the closeness of the protein packing suggests that dipole charges will generate significant attractive forces between adjacent molecules. For example, where two spheroids the size of Hb are almost touching and where their distance apart (surface to surface) is of the order 1/10th of the radius of the spheroid, then the interaction energy has been shown to be of the order 10^{-12} ergs or greater [14]. This is 0.6 electron volts, which is huge when compared to the average kinetic energy at body temperature. This inter-dipole attractive force between adjacent haemoglobin molecules is of the same order of magnitude as a hydrogen bond. All of the above suggests that the dipole attractions between adjacent haemoglobins are significant, and will play a crucial role in orienting and spacing the haemoglobins within their liquid lattice structure.

5. EFFECT OF THE CHARGE CHANGE WHICH OCCURS ON SUBSTRATION OF HAEMOGLOBIN

A haemoglobin molecule with 3 of its 4 haem binding sites occupied by oxygen will carry a significant additional dipole compared to a fully saturated or fully desaturated haemoglobin molecule. The additional dipole is due to the charge change which occurs on substration, and is a consequence of the hydrogen ion exchange which accompanies oxygen substration or desubstration (the Bohr shift). Additionally, the uptake and unloading of carbon dioxide will also generate a charge change, as will the other temporary liaisons between the Hb and ATP and 2,3DPG.

It is apparent that from the highest degree of saturation achieved (98%) to the lowest degree of desaturation in venous blood, there will always be enough haemoglobins partially oxygenated or deoxygenated so that each such incompletely oxygenated haemoglobin is directly adjacent to one or more similarly partly oxygenated haemoglobins. Since they are held in place within a quais-lattice, but are all free to make rotational jumps, we believe that these haemoglobins in particular, each carrying significant additional dipoles, will orient relative to adjacent haemoglobins so as to form dipole-dipole bonds. Additionally these dipole generated orientations will have the consequence of placing a vacant haem site in one haemoglobin closer to a filled haem site on the adjacent Hb for a significantly longer time than would be the average time due to random orientations.

Of course, Brownian rotational diffusion will tend to distribute the dipoles about their lowest energy configuration as demanded by the dipoles. This lowest energy configuration between two adjacent haemoglobins will occur when the two dipoles are aligned so that an empty haem in a partly deoxygenated molecule is closer to an occupied haem in the next adjacent haemoglobin, particularly if it is also partially deoxygenated.

It can be shown that adjacent haemoglobins are only separated (average closest surface separations) by 2.5Angstrom. The distance separating them is thus of the order $1/10^{th}$ of the radius of the particles, where the interaction energies are certainly large.

6. EFFECT OF LOADING OR UNLOADING OXYGEN

As outlined above, a haemoglobin molecule with three of its four sites occupied with oxygen will have a significant additional dipole due to the consequent additional dipole asymmetry (itself consequent on the inverse of the Bohr effect). Thus when the fourth site is occupied, this asymmetry disappears as does the additional dipole. Because of the consequently amended dipole interaction energy between these adjacent haemoglobins, the haemoglobin just oxygenated will make a rotational jump into a new position of lowest energy configuration associated with the new dipole which it carries. Thus, each oxygenation or deoxygenation will generate rotational movement among the haemoglobins which will of course tend to move the freshly loaded oxygen further down its concentration gradient and of course place it in a new region ready to make another jump. Additionally, the new orientation in the E-M field of the cell membrane and the surrounding dipoles will also tend to amend the charge distribution within the haemoglobin (the induced dipole), and this in turn will modify the binding affinity of oxygen for the haem site (partially through the Bohr effect).

7. PRESENCE OF ELECTRIC FIELDS AT THE RED CELL MEMBRANE

An electromagnetic field will inevitably produce a force upon a charged particle, or it will orient a dipole. The cell membrane generates large fields, since the membrane resting potential is of the order 100 mV across a membrane with a thickness of the order 10nm. This is a field of 10 million volts/metre. While it is very rapidly attenuated with distance, a significant E-M field extends for a small distance into the body of the cell. Within this layer adjacent to the inner surface of the cell membrane, the haemoglobin dipoles will be oriented primarily by the cell membrane E-M field. As unloading of oxygen by the red cell proceeds at the inner surface of the red cell membrane, the change in the dipole associated with desubstration, will cause the haemoglobin to flip around in a rotational movement, which will then place another oxygenated haem closer to the membrane. Thus the rotation has the effect of presenting the next oxygenated haem into the surface region where unloading will be further encouraged.

The same mechanism which encourages unloading on the inside surface of an oxygenated erythrocyte will also promote loading of oxygen within lung capillaries. Here the gradient of free oxygen is highest on the side of the haemoglobin presented towards the red cell membrane, so loading of oxygen will be onto the outer side of the haemoglobins laying immediately inside the erythrocyte membrane. As soon as the oxygen is bound, the resulting charge change consequent on this substration will encourage a rotational movement which will push the oxygen towards the interior of the erythrocyte, while presenting a new empty haem site to the red cell membrane side.

8. CHANGING SHAPE OF THE RED CELL

As the red cell passes through narrow capillaries, it undergoes shape and, temporarily, volume changes. The volume changes will be promptly restored by the membrane pumps, but the consequence of the shape changes will be to stir the cell contents and thus shuffle the intracellular haemoglobins. This shuffling may play an important role in diminishing the oxygen gradients between the centre and the periphery of the erythrocyte, and thus in enhancing the steepness of the gradient across the red cell membrane, with a consequent increase in the flux of oxygen across the membrane.

9. REFERENCES

1. Wyman,J Facilitated diffusion and the possible role of myoglobin as a transport molecule. *J.Biol.Chem.* **241** 115-121 (1966)
2. Murray J D, On the molecular mechanism of facilitated oxygen diffusion b y haemoglobin and myoglobin. Proc. R. Soc. Lond. B. Biol. Sci. , **178**, 95-110. (1971)
3. Wittenburg, JB The molecular mechanism of haemoglobin facilitated oxygen diffusion. *J. Biol.Chem.* **241** 104-114 (1966)
4. Wittenburg JB and Wittenburg BA Facilitated oxygen diffusion by oxygen carriers. In *Oxygen and Living Processes* Ed D.Gilbert Springer-Verlag New York pp 177-199 (1981)
5. Banks,DS and Fradin,Cecile, Anomalous diffusion of proteins due to molecular crowding. *Biophys.J* **104** 1078-1084. (2005)
6. Arrio-Dupont,Martine, Cribier, S. Foucault, G. Devaux, PF and D'Albis, A. Diffusion of fluorescently labelled macromolecules in cultured muscle cells. *Biophys.J* **70** 2327-2332 (1966)
7. Jacobson,K and Wojcieszyn, The translational mobility of substances within the cytoplasmic matrix *Proc. Natl Acad.Sci.USA* **81,** 6764-6751 (1984)
8. Wittenburg JB and Wittenburg BA Myoglobin function reassessed. *J.Exp.Biol* **206** 2011- 2020 (2003)
9. Bragg WL, Howells ER and Perutz MF Arrangement of polypeptide chains in horse methaemoglobin. Acta Cryst. **5**, 135-141 (1952)
10. Ingram DJE Gibson JF & Perutz,MF Orientation of the four haem groups in haemoglobin. Nature, **178,** 905-910, (1956)
11. Preston B, Obrink, B and Laurent, T.C. Rotational diffusion coefficient of albumin within a polymer matrix. *Eur.J.Biochem* **33;** 401-406. (1973)
12. McCabe M and Maguire D. In *Advances in Experimental Medicine and Biology* (2004)
13. Setlow, R.B. and Pollard, E.C. Dipole-dipole interactions, in *Molecular Biophysics; Addison-Wesley Publishing Co Inc Mass.* USA pp165-180 (1962)
14. Vold, MJ, Van der Waal's attraction between anisometric particles. *J.Coll. Sci.* **9:** 451-459 (1954).

SIMULTANEOUS MEASUREMENT OF pO_2 AND PERFUSION IN THE RABBIT KIDNEY *IN VIVO*

Paul M. O'Connor, Warwick P. Anderson, Michelle M. Kett, and Roger G. Evans*

Abstract: Recently, a combined probe has been developed capable of simultaneous measurement of local tissue pO_2 (fluorescence oximetry) and microvascular perfusion (laser Doppler flux) within the same local region. The aim of the current study was to test the utility of these combined probes to measure pO_2 and perfusion in the kidney. Studies were performed in anesthetized, artificially ventilated rabbits (n=7). Baseline measurements of renal medullary perfusion and pO_2 obtained using combined probes (537±110 units & 28.7±6.1mmHg, respectively) were indistinguishable from those obtained using independent probes (435±102 units & 26.9±6.4mmHg). Baseline measurements of renal cortical pO_2 were also similar between combined (9.7±1.6mmHg) and independent probes (9.5±2.3mmHg). Baseline levels of cortical perfusion however, were significantly greater when measured using independent probes (1130±114units) compared to combined probes (622±59units; $P<0.02$). Relative changes in perfusion and pO_2 resulting from graded stimulation of the renal nerves were not significantly different when measured using combined probes to those obtained using independent probes. We conclude that combined probes are equally suitable to independent probes for tissue pO_2 and microvascular perfusion measurement in the kidney. Our results raise some concerns regarding the accuracy of these OxyLite fluorescence probes for pO_2 measurement in the kidney, particularly within the renal cortex.

1. INTRODUCTION

Measurement of renal tissue pO_2 and perfusion has previously required separate specialized probes[1, 2]. Because both renal tissue pO_2 and microvascular perfusion are

* Paul M. O'Connor, Warwick P. Anderson, Michelle M. Kett and Roger G. Evans, Department of Physiology, Building 13F, Monash University, Victoria 3800, Australia.

regionally heterogeneous[2, 3], a limitation of this approach is that these parameters must be measured in different areas of tissue. Recently, a combined probe has been developed, for simultaneous measurement of tissue pO_2 (fluorescence oximetry) and microvascular perfusion (laser Doppler flux) in the same local tissue region. These combined probes have been used successfully in other tissues[4, 5], but have not yet been tested in the kidney, an organ with a remarkably high level of perfusion per unit of tissue volume[6]. Therefore, in the current study we compared measurements of tissue pO_2 and laser Doppler flux obtained using combined probes, to those of more conventional independent measurements of tissue pO_2 and perfusion, in the rabbit kidney *in vivo*. In order to compare the responsiveness of the different probe types, we altered renal microvascular perfusion and tissue pO_2 by graded electrical stimulation of the renal nerves.

2. METHODS

Experiments were performed on 7 male New Zealand White rabbits (2.96 ± 0.12 kg), in accordance with the Australian Code of Practice for the Care and Use of Animals for Scientific Purposes and were approved in advance by the Animal Ethics Committee of the Department of Physiology, Monash University.

2.1. Surgical preparation

Rabbits were anesthetised with pentobarbital (90-150 mg plus 30-50 mg/h; Nembutal; Merial Australia Pty Ltd, NSW, Australia), intubated, and artificially ventilated. Compound sodium lactate (Baxter Health Care, Toongabbie, NSW, Australia) was infused intravenously (0.18 ml $kg^{-1}min^{-1}$) to replace fluid loss. Esophageal temperature was maintained at ~37.5 °C. The left kidney was placed in a micropuncture cup for stability and a transit-time ultrasound flow probe was placed around the left renal artery for measurement of total renal blood flow (TRBF; Type 2SB, Transonic Systems Inc, Ithaca, NY, USA). Renal nerves were isolated and placed on a set of hooked electrodes[7, 8]. Ventilation rate was then adjusted to set arterial pO_2 between 90 and 110 mmHg. Following instrumentation, the maintenance infusion was replaced with a solution containing four parts compound sodium lactate to 1 part of a 10% polygeline/electrolyte solution (Haemacell; Hoechst, Victoria, Australia). Experimental manipulations commenced after a 90 min equilibration period. At the conclusion of the experiment animals were killed with an intravenous overdose of pentobarbital.

2.2. Measurement of pO_2 and microvascular perfusion

Independent measurements of medullary laser Doppler flux (MLDF) and medullary pO_2 (MpO_2) were obtained using a needle laser Doppler flow probe (MNP110XP, tip diameter = 500 μm, Oxford Optronix, Oxford, England) and a single fiber pO_2 probe (BF/OT, tip diameter = 350 μm Oxford Optronix), hereafter referred to as 'independent probes'. Independent measurements of cortical pO_2 (CpO_2) and cortical laser Doppler flux (CLDF), were obtained using a surface laser Doppler probe (MSP310XP, Oxford Optronix) and a single fiber pO_2 probe (BF/OT, Oxford Optronix), respectively. Simultaneous measurements of MpO_2 and MLDF, or of CpO_2 and CLDF, in the same local tissue region, were obtained using a triple fiber pO_2/LDF probe (BF/OT/2.0NS, tip diameter = 450 μm Oxford Optronix), hereafter referred to as a 'combined probe'. The MNP110XP, BF/OT and BF/OT/2.0NS probes were inserted into the kidney using a micromanipulator so that their tips lay ~ 9 mm apart, either 10-12 mm (medulla) or 2 mm (cortex) below the mid-region of the lateral surface of the kidney. The surface laser Doppler probe was placed on the ventral surface of the mid region of the kidney. For each animal, the insertion points of the probes, and thus their locations within the kidney, were randomized using a Latin square design.

2.3. Experimental protocol

The renal nerves were electrically stimulated as previously described[7]. Probes were initially positioned in either the medulla or cortex as described above. The renal nerves were then stimulated at frequencies of 0.5, 1, 2 and 4 Hz in random order (8 V, 2 ms duration, 3 min trains). The probes were then positioned in the alternate region (either cortex or medulla) and the stimulation protocol repeated. Each stimulus train was preceded by a 5 min control period, and followed by a 3-5 min recovery period.

2.4. Measurement of hemodynamic variables

Arterial pressure was measured via an ear artery catheter. Laser Doppler flux and pO_2 probes were connected to a laser Doppler flowmeter/tissue oximetry system (OxyLite/Oxyflo, Oxford Optronix). Analog signals were digitized and recorded as previously described[6, 7]. Baseline levels of CLDF and MLDF, obtained during brief renal artery occlusion, were subtracted from experimental data before subsequent analysis. The position of each of the probes was also checked at post-mortem.

2.5. Statistical analysis

Data are presented as mean ± SEM. Paired t-tests were used to test whether baseline levels differed between combined and independent measurements of tissue pO_2 and laser Doppler flux. Responses to renal nerve stimulation were calculated from the average values during each stimulation period and are presented as percentage changes from the preceding control periods. Two-way analysis of variance was used to determine whether the frequency-dependent effects of renal nerve stimulation were affected by the probe type used to aquire the measurements (independent or combined). Model II regression analysis was used to determine the relationships between combined probe and independent probe measurements of tissue pO_2 and laser Doppler perfusion[9].

3. RESULTS

Stimulation of the renal nerves resulted in immediate and marked reductions in TRBF. The response was frequency dependent ($P<0.001$). There were also frequency dependent reductions in CpO_2, MpO_2, CLDF and MLDF. These data were subjected to analysis of variance, which showed that relative changes in both CpO_2 and MpO_2 obtained using a combined probe were indistinguishable from those obtained using the independent pO_2 probes. Similarly, relative changes in MLDF using combined probes were indistinguishable from those obtained using an independent needle LDF probe. There was a tendency for responses of CLDF to renal nerve stimulation to be greater when measured by surface probe compared to the combined probe (P = 0.08).

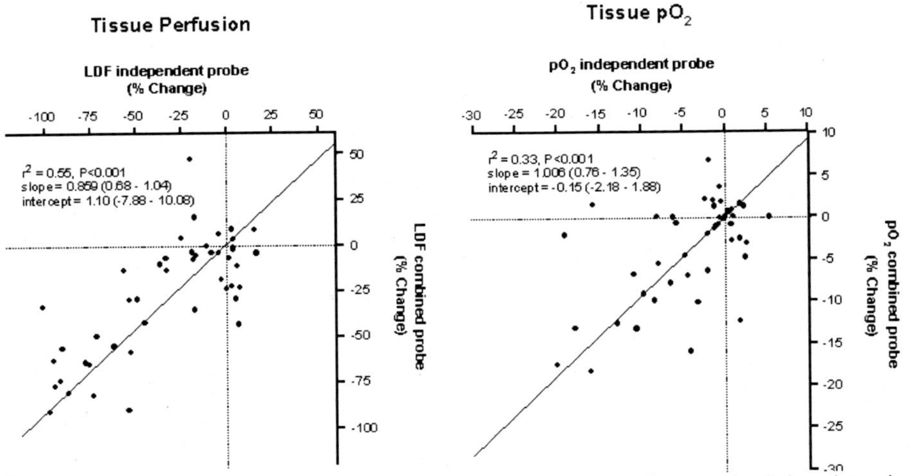

Figure 1. Scatter-plot of changes in tissue LDF and pO_2 in response to renal nerve stimulation, determined using independent and combined probes in the rabbit kidney. The line of best fit was obtained using the least-products method (model II regression analysis[8]).

Figure 1 shows scatter-plots of percentage changes in pO_2 and LDF (cortex and medulla combined) in response to renal nerve stimulation. Some variability in responses was observed, but strong linear relationships were present, with the 95% confidence intervals of the slopes including unity, and those of the x-intercepts including zero.

Table 1. Baseline levels of tissue pO_2 and laser Doppler flux in the rabbit kidney.

	Independent Probe	Combined Probe	P
CLDF (units)	1130 ± 114	622 ± 59	0.02
MLDF (units)	435 ± 102	537 ± 110	0.5
CpO_2 (mmHg)	9.5 ± 2.3	9.7 ± 1.6	0.9
MpO_2 (mmHg)	26.9 ± 6.4	28.7 ± 6.1	0.9

CLDF, cortical laser Doppler flux; MLDF, medullary laser Doppler flux, CpO_2, cortical tissue pO_2; MpO_2, medullary tissue pO_2. P values were derived from unpaired t-tests comparing baseline values obtained using independent and combined probes (n = 7).

Baseline levels of mean arterial pressure (70 ± 3 mmHg), heart rate (245 ± 6 beats min^{-1}) and TRBF (26.9 ± 4.2 ml min^{-1}) remained relatively stable throughout the course of the experiment. Arterial pO_2 averaged 92 ± 12 mmHg. Tissue temperature within the renal medulla was 36.6 ± 0.1° C while cortical tissue was cooler, averaging 33.4 ± 0.2° C, during control periods. Baseline levels of MLDF, CpO_2 and MpO_2 were similar when measured using combined or single probes (Table 1). Baseline levels of CLDF, however, were significantly greater when measured using an independent probe compared to measurements obtained using a combined probe (Table 1).

4. DISCUSSION

The major finding of this study is that relative changes in tissue pO_2 and microvascular perfusion measured with combined pO_2/laser Doppler probes are similar to those obtained using independent probes. Analysis of variance showed that both in the cortex and medulla, and both for pO_2 and LDF, responses to renal nerve stimulation observed using combined probes were not significantly different from those observed using independent probes. In addition, model II regression analysis showed that for both tissue pO_2 and LDF, the relative changes in response to renal nerve stimulation observed using combined and independent probes could be described by straight lines whose slopes did not differ from unity, and whose intercepts did not differ from zero.

4.1. Measurement of tissue perfusion using combined and independent probes

Baseline CLDF was less when measured using a combined probe compared to the independent probe. As baseline levels of MLDF measured by combined and independent probes were similar, it seems unlikely that the lower CLDF with the combined probe is related to the positioning of the laser Doppler fibers behind the oxygen sensing matrix in these probes. Instead, the lower baseline flux appears to be specifically associated with some aspect of LDF measurement specifically within cortical tissue.

One explanation for the difference in baseline CLDF between the two probe types in the current study may be that combined and independent measurements of LDF came from slightly different regions of the renal cortex. Palm *et al.* have recently shown a steep gradient for LDF within the renal cortex of rats, LDF being greatest within 1 mm of the cortical surface[2]. The higher LDF in this region could be due to the greater hematocrit in the superficial cortex[10]. In the current study, independent measurements of CLDF were obtained using a surface laser Doppler probe. These probes were placed on the renal surface and have an estimated measurement depth of 0.3 to 1.5 mm, depending on the type of tissue in which they are used. Thus, it is likely that these probes would have provided measurements from within the first mm of tissue below the surface of the kidney, the region of tissue in which flux is greatest[2]. In contrast, combined probes were positioned so that the probe tip lay 1-2 mm below the surface of the kidney. Even after accounting for the fact that the optical fibers that measure LDF are recessed ~500 μm behind the probe tip, measurements of flux from these combined probes would have come from tissue up to 2 mm below the renal surface, where CLDF might be less than in the superficial cortex.

An alternative explanation for the relatively low CLDF measured with the combined probe may be damage to the cortical microvasculature in the field in which measurements were obtained. Independent measurements of CLDF were made using a surface probe that should not have caused damage to the underlying microvascular, whilst combined measurements of CLDF were made using a probe inserted ~2 mm into the renal parenchyma.

4.2. Utility of 'OxyLite' fluorescence probes for measurement of renal tissue pO_2

Although the OxyLite system has been successfully used by others to measure tissue pO_2 levels in the spleen[4], thymus[4] and brain[5], the ability of these fluorescence probes to measure tissue pO_2 in the kidney had not yet been tested. Values reported for medullary pO_2 in the literature using other techniques range between 5-30mmHg[1, 3, 2, 11] In the current study, mean values of MpO_2 obtained using both combined and independent probes were in the range of 25-30mmHg, thus, it would appear that within renal medullary tissue, these fluorescence probes are capable of accurate pO_2 measurement. This approach also offers some advantages, over Clark O_2 electrodes, for the measurement of tissue pO_2 within the kidney. OxyLite probes do not consume O_2, eliminating the need to move the probe, once it is positioned, to avoid creating local tissue O_2 gradients[12]. While micro-oxygen electrodes largely overcome this issue, creating minimal oxygen gradients, miniaturization also limits the catchment area in which measurements are made. As these optical probes average pO_2 over a larger area of tissue than micro-electrodes, they may produce a better overall indication of pO_2 in tissues with heterogeneous pO_2 profiles, such as the kidney[3,4].

In the current study, values of CpO_2 obtained using both combined and independent probes in rabbits were ~10 mmHg, considerably lower than previously published values using other techniques in rats (30-40 mmHg)[2, 11, 13] . Because we did not directly compare these values with those obtained with an alternative technique such as a Clark electrode, the extent to which these relatively low values are artifactual remains unknown. Both the independent and combined optodes used to measure CpO_2 in the current study are much larger than Clark-type electrodes typically used for pO_2 measurement within the kidney, which generally have a tip diameter of less than 20 μm[11]. Thus the possibility exists that the OxyLite fluorescence probes used in the current study cause tissue damage in the renal cortex, which reduces pO_2 in the surrounding tissue. A second possibility is that the CpO_2 of the animals used in the current study was low, and that CpO_2 measurements obtained using OxyLite fluorescence probes were accurate. In support of this possibility, cortical tissue temperature in the current study was lower than medullary tissue temperature. The low temperature of the renal cortex probably resulted from exposure of the kidney surface to the ambient temperature (~20 °C) of the laboratory environment. Cooling of the cortical tissue may have resulted in constriction of the cortical microvasculature and thus reduced cortical O_2 delivery[14].

5. SUMMARY AND CONCLUSIONS

In summary, relative changes in tissue pO_2 and microvascular perfusion obtained using combined pO_2/laser Doppler probes are virtually indistinguishable from those obtained using independent probes. We conclude that combined probes are equally suitable to independent probes for tissue pO_2 and microvascular perfusion measurements in the *in vivo* kidney. Indeed, given that the combined probes provide measurements of tissue pO_2 and microvascular perfusion in the same local region of tissue, these probes may be preferred in certain studies. Nevertheless, our results do raise some concerns regarding the accuracy of absolute measurements of renal tissue pO_2 using these fluorescence optodes, particularly within the renal cortex.

6. REFERENCES

1. M. Brezis, S.N. Heyman, and F.H. Epstein, Determinants of intrarenal oxygenation. II. Hemodynamic effects. *Am. J. Physiol.* **267**(6 Pt 2), F1063-F1068 (1994).
2. F. Palm, J. Cederberg, P. Hansell, P. Liss, and P.-O. Carlsson, Reactive oxygen species cause diabetes-induced decrease in renal oxygen tension. *Diabetologia* **46**(8), 1153-1160 (2003).
3. D.W. Lubbers, and H. Baumgartl, Heterogeneities and profiles of oxygen pressure in brain and kidney as examples of the pO_2 distribution in the living tissue. *Kid. Int.* **51**(2), 372-380 (1997).
4. R.D. Braun, J.L Lanzen, S.A. Snyder, and M.W. Dewhirst, Comparison of tumor and normal tissue oxygen tension measurements using OxyLite or microelectrodes in rodents. *Am. J. Physiol.* **280**(6), H2533-H2544 (2001).
5. C.I. Nwaigwe, M.A. Roche, O. Grinberg, and J.F. Dunn, Brain tissue and sagittal sinus pO_2 measurements using the lifetimes of oxygen-quenched luminescence of a ruthenium compound. *Adv. Exp. Med. Biol.* **530**, 101-111 (2003).
6. J. Meier, A. Pape, M. Kleen, J. Hutter, G. Kemming, and O. Habler, Regional blood flow during hyperoxic haemodilution. *Clin. Physiol. Funct. Imaging* **25**, 158-165 (2005).
7. B.L. Leonard, S.C. Malpas, K.M. Denton, A.C. Madden, and R.G. Evans, Differential control of intrarenal blood flow during reflex increases in sympathetic nerve activity. *Am. J. Physiol.* **280**(1), R62-R68 (2001).
8. S.J. Guild, G.A. Eppel, S.C. Malpas, N.W. Rajapakse, A. Stewart, and R.G. Evans, Regional responsiveness of renal perfusion to activation of the renal nerves. *Am. J. Physiol.* **283**(5), R1177-R1186 (2002).
9. J. Ludbrook, Comparing methods of measurement. *Clin. Exp. Pharmacol. Physiol.* **24**(2), 193-203 (1997).
10. J.R. Pappenheimer, and W.B. Kinter, Hematocrit ratio of blood within mammalian kidney and its significance for renal hemodynamics. *Am. J. Physiol.* **185**(2), 377-390 (1956).
11. P. Liss, A. Nygren, N.P. Revsbech, and H.R. Ulfendahl, Measurements of oxygen tension in the rat kidney after contrast media using an oxygen microelectrode with a guard cathode. *Adv. Exp. Med. Biol.* **411**, 569-576 (1997).
12. J.R. Griffiths, and S.P. Robinson, The OxyLite: a fibre-optic oxygen sensor. *Br. J. Radiol.* **72**, 627-630 (1999).
13. W.J. Welch, H. Baumgartl, D. Lubbers, and C.S. Wilcox, Nephron pO_2 and renal oxygen usage in the hypertensive rat kidney. *Kid. Int.* **59**(1), 230-237 (2001).
14. M.N. Levy, and E.S. Imperial, Oxygen shunting in renal cortical and medullary capillaries. *Am. J. Physiol.* **200**(1), 159-162 (1961).

PSEUDOGENES AND THE ELECTRON TRANSPORT CHAIN

H. M. Oey*, D. J. Maguire*, and M. McCabe*

Abstract: With the advent of easy access to the human genome sequence, molecular biology techniques to target respirome-specific genes have begun to be exploited in the study of human disorders and in particular human cancers. In some recent publications it would appear that some investigators have inappropriately targeted pseudogenes rather than functional genes. The high transcription level and generally small size of many of the genes in the respirome make them prone to duplications in the form of processed pseudogenes within the human genome. Such genes can be challenging to analyse using standard molecular genetics approaches. In this presentation, we offer an analysis of pseudogenes that have been identified to have significant homology with some elements of the respirome. Other sequence elements such as Alu repeats, which present similar research obstacles, are also discussed.

1. INTRODUCTION

Pseudogenes are a well known genomic feature of virtually all types of living organisms. An exact definition of a pseudogene is hard to state, however, in order to be considered a pseudogene, a sequence should be homologous to a functional parent sequence, a gene, and it should have lost the function of that parent gene. Pseudogenes are therefore mostly non-functional relics under minimal or no selective pressure. They are free to accumulate mutations and therefore usually only have limited homology to the functional paralogous gene.

In mammals, there are two main types of pseudogenes; the processed pseudo genes and the duplicated pseudogenes.[1, 2] Duplicated pseudogenes are formed through recombination events or unequal crossing over of chromosomal DNA. In order to become a pseudogene, the duplicated gene must be disabled either transcriptionally because of missing or damaged promoters or transcription initiation signals, or translationally due to mutations causing premature stop-codons, frameshift mutations, etc.[3-6]

* School of Biomolecular and Biomedical Science, Griffith University, Queensland 4111, Australia.

Processed pseudogenes are formed via an entirely different mechanism; it is believed that these sequences originate from processed messenger RNA transcripts (mRNA) that are reverse transcribed and reinserted back into the genome. These sequences are therefore intronless copies of the paralogous gene and often found with remnants of post-transcriptionally added poly-adenine (polyA) tails at their 3' ends.[7, 8]

The mechanism involved in the reverse transcription event is believed to be supplied by the retrotransposon long interspersed element 1 (LINE1). LINE1 has two open reading frames, one of which encodes a reverse transcriptase that facilitates the reverse transcription and reinsertion back into the genome of LINE1s own transcript.[9-11] LINE1 however, is not the only gene to make use of this mechanism, and surprisingly it is not even the most successful at it. Another transposon, the Alu repeat, a member of the short interspersed repeats (SINE), makes use of the same mechanism and has been found to have more than 1.1×10^6 copies of itself spread throughout the human genome,[12] a significantly greater number than LINE1 with 5.2×10^5 copies.[13]

Processed pseudogenes are believed to rely on the same mechanism. It has also been shown that these pseudogenes had their greatest insertion rates concurrently with the Alu repeats, suggesting a shared mechanism of retroposing. In mice however, which do not have Alu repeats, the insertion rates appear to follow that of LINE1.[6, 14]

Since processed pseudogenes are formed from RNA transcripts, it is perhaps not surprising that genes transcribed at high rates also tend to have greater numbers of processed pseudogenes.[3] It has also been shown that shorter genes that are expressed in a wide variety of tissues also have more pseudogenes. The latter feature is probably caused by a requirement for the gene to be expressed in the germ-line in order to become fixed in the population.[3] Most housekeeping genes, including the genes involved with metabolic pathways, satisfy these parameters.

Pseudogenes are generally considered to be junk DNA; non-functional DNA sequences that are not under any form of selective pressure and thus free to accumulate mutations and eventually erode away. While this assertion is generally true, there are exceptions; multiple examples of pseudogenes that are transcribed now exist[15, 16] and there are even suggestions of some that are translated.[17-19] Furthermore, some pseudogenes have been shown to be involved in the regulation of the expression of their parent genes, which complicates the definition of a pseudogene.[20, 21] Other more theoretical functions of pseudogenes have also been proposed. These especially included the potential for pseudogenes to be transcribed in antisense and form RNA duplexes with the original mRNA transcript, thus affecting their expression.[18, 22, 23]

Some of the most fundamental proteins in eukaryote cells are the ones involved with the electron transport chain (ETC) and ATP production. These proteins, while mostly being nuclear encoded, reside at the inner mitochondrial membrane.[24, 25] While most of these proteins are thought to have been once encoded by the mitochondrial genome itself, and some still are, the majority are now encoded by the nuclear genome.[26, 27] Most of the genes in this group fit the parameters for genes likely to give rise to processed pseudogenes quite well, and not surprisingly, most of them have done just that. We provide an overview of the ETC genes and their pseudogenes.

2. METHODOLOGY

The mRNA sequences for the genes of the ETC (table 1), were retrieved from ENSEMBL (www.ensembl.org). The corresponding gene sequence was retrieved from NCBI, human genome version 35.1. Paralogous DNA sequences were found using "cross-species MegaBLAST"[28] (www.ncbi.nih.gov) allowing for more diverged sequences to be aligned. The searches were performed without filters for repetitive DNA unless such a filter was required to obtain a search result; no other advanced settings were defined. Aligned sequences were matched with corresponding entries at Pseudogene.org (www.pseudogene.org). Only aligned sequences longer than 200 base pairs (bp) were included, no minimum score-threshold was otherwise defined. Pseudogenes with discontinuous alignments were considered to be single processed pseudogenes unless the gaps matched the splice sites of the parent gene.

3. RESULTS

A total of 73 nuclear encoded genes known to be involved with the electron transport chain were analysed for pseudogenes, both from entries at Pseudogene.org (www.pseudogene.org) and through BLAST-searches[28] of the human genome. It was shown that at least 51 genes out of the 73 genes analysed did have one or more pseudogenes, and about half of these genes had more than one pseudogene (table 1). It is also a trend that the genes found to have pseudogenes associated with them also have short mRNAs. Only four duplicated pseudogenes, all related to the same original gene were found in this survey. ETC complex I is the complex with the greatest number of genes associated with it, and it also has the smallest fraction of genes with pseudogenes.

Table 1. Genes and pseudogenes of the ETC.

ETC Complex	Gene name	Gene length	mRNA length	Count of exons	Count of pseudogenes
ETC Complex I	NDUFA1	4885	479	3	2
	NDUFA2	2283	600	3	1
	NDUFA3	4109	360	4	5
	NDUFA4	6999	823	4	25
	NDUFA5	16876	1550	5	14
	NDUFA6	5359	1202	3	None
	NDUFA7	10023	531	4	None
	NDUFA8	15762	859	4	None
	NDUFA9	38117	1334	11	1
	NDUFA10	64643	1557	10	None
	NDUFA11	9343	819	4	None
	NDUFAB1	15305	804	5	1
	NDUFB1	5687	433	3	2
	NDUFB2	9966	509	4	1
	NDUFB3	13932	693	3	5
	NDUFB4	6047	541	3	11
	NDUFB5	19714	1088	6	2
	NDUFB6	19660	862	4	1
	NDUFB7	5997	570	3	None
	NDUFB8	6140	686	5	4
	NDUFB9	10864	740	4	4
	NDUFB10	2458	736	3	2
	NDUFB11	2819	953	3	1
	NDUFC1	5867	418	4	1
	NDUFC2	11536	2159	3	None
	NDUFS1	36385	3417	19	None
	NDUFS2	15081	2061	15	None
	NDUFS3	5483	899	7	None
	NDUFS4	122684	668	5	None

Table 1. (Continued)

	NDUFS5	8282	540	3	7
	NDUFS6	14660	750	4	1
	NDUFS7	11677	787	8	None
	NDUFS8	6006	779	7	None
	NDUFV1	5599	1566	10	None
	NDUFV2	31610	827	8	2
	NDUFV3	16396	2151	4	None
ETC Complex II	SDHA[a]	38346	2277	15	4
	SDHB	35394	1100	8	1
	SDHC	48808	1315	6	5
	SDHD	8896	1313	4	8
ETC Complex III	CYC1	2430	1273	6	None
	HSPC051	3045	934	2	1
	QP-C	817	388	2	5
	UQCR	8261	1305	3	1
	UQCRC1	10667	1636	13	None
	UQCRC2	30060	1674	14	3
	UQCRFS1	5948	1203	2	3
	UQCRH	13041	515	4	6
ETC Complex IV	COX4I1	7413	802	5	2
	COX5A	17754	645	5	4
	COX5B	2137	523	4	8
	COX6A1	2626	548	3	7
	COX6B1	10530	578	4	7
	COX6C	15524	444	4	30
	COX7A1	1948	783	4	None
	COX7B	5819	456	3	10
	COX7C	2800	448	3	6
	COX8A	1937	521	2	None
ETC Complex V	ATP5A1	20090	1950	13	9
	ATP5B	7894	1857	10	2
	ATP5C1	19672	1162	10	1
	ATP5D	3076	1005	4	None
	ATP5E	3690	498	3	2
	ATP5G1[b]	3086	663	5	15
	ATP5G2[b]	11481	1094	5	14
	ATP5G3	3847	1014	5	3
	ATP5H	8120	628	6	4
	ATP5I	1898	368	4	None
	ATP5J	11175	1303	4	5
	ATP5L	8459	1343	3	12
	ATP5O	12402	815	7	2
	ATP5S	13622	1567	5	None

[a] The four pseudogenes for SDHA are all duplicated pseudogenes.
[b] The genes ATP5GI and ATP5G2 have homologous coding sequences and therefore align with the same pseudosequences.

4. DISCUSSION

It is evident that pseudogene formation is a common feature of the ETC genes. There are a number of potential reasons for this. The most important reasons are probably the fact that the ETC genes are housekeeping genes, and therefore are constantly transcribed at high levels and with little tissue specificity, consistent with genes that are likely to form pseudogenes.[3] It is also a trend that ETC genes with pseudogenes are short genes with few exons, also consistent with previously published data.[3]

The ETC genes seem to almost exclusively form processed pseudogenes rather than duplicated pseudogenes. The reason for this preference is probably that duplicated pseudogenes are formed through an entirely different process than processed pseudogenes. The recombination events that form duplicated pseudogenes should not be significantly affected by the transcriptional nature of a gene, and processed pseudogene formation is therefore more common among these genes.

Some, but not all, of the ETC genes originate from the mitochondrial genome. While some of these mitochondrial genes persist in this genome, most have now been transferred to the nuclear genome where they have taken on eukaryotic characteristics. Exactly which genes this includes is not completely understood.[27, 29] The large difference seen between the different genes and complexes of the ETC for processed pseudogene formation may be due to these different origins. The age of a gene in the nuclear genome may dictate how many processed pseudogenes they have had time to form and a mitochondrial versus a genomic origin of genes may also have had an effect on pseudogene formation.

It seems that Complex I has a lower percentage of genes with processed pseudogenes than the other complexes. Many of the genes encoding the proteins of this complex are believed to be of a nuclear origin since the mitochondrial genome is not believed to have supported the high number of proteins associated with it.[30-33] Perhaps the greater number of genes from a non-mitochondrial origin is one reason for this effect.

A number of the entries in table 1 have Alu repeats within them These sequences are sometimes mistakenly considered to be duplicated pseudogenes in some entries at pseudogene.org,[6, 16] and other pseudogene datasets.[5, 34] However, the presence of intervening retroposons, such as Alu repeats or LINE1, and lack of splice-sites matching the paralogous parent gene, suggest these are actually processed pseudogenes.

Some entries in table 1 are identical for different genes. This is probably due to a common origin or exchange of genetic material between these genes. This survey primarily looks at genomic regions with high similarity to each gene of the ETC, it does not try to identify the origin of each such pseudogene sequence. Therefore, some sequences are strictly speaking not pseudogenes of the genes in table 1, but rather just homologous sequences originating from a paralogous gene.

It is evident that processed pseudogenes may pose significant problems when using various methods common in molecular genetics research. These especially include protocols that involve reverse transcription of mRNA, such as northern blots. A processed pseudogene is an exact replica of an mRNA transcript, incomplete removal of genomic DNA may therefore cause false positive results when amplifying cDNA from these genes.

Poorly designed primers may also cause amplification of processed pseudogenes when the target is the genomically encoded gene itself. Use of primers that bind intronic DNA or at exon intron boundaries mediates this problem. However, if the gene has a duplicated pseudogene the problem will persist regardless of where the primers bind since such pseudogenes also have the same intronic sequences. Duplicated pseudogenes are well known to cause such problems.[35, 36]

An overview of the genes and pseudogenes of the respirome is a valuable resource for molecular genetics research into the genes of the ETC, and may prove to limit problems surrounding primer design and non-specific amplification during PCR and reverse transcription of mRNA.

5. REFERENCES

1. E. F. Vanin. Processed pseudogenes: Characteristics and evolution. *Annu. Rev. Genet.* **19,** 253-272 (1985).
2. A. J. Mighell, N. R. Smith, P. A. Robinson, and A. F. Markham. Vertebrate pseudogenes. *FEBS Lett.* **468,** 109-114 (2000).
3. I. Goncalves, L. Duret and D. Mouchiroud. Nature and structure of human genes that generate retropseudogenes. *Genome Res.* **10**(5), 672-678 (2000).
4. V. E. Prince and F.B. Pickett. Splitting pairs: the diverging fates of duplicated genes. *Nat Rev Genet.* **3**(11),827-37 (2002).
5. D. Torrents, M. Suyama, E. Zdobnov and P. Bork. A genome-wide survey of human pseudogenes. *Genome Res.* **13**(12),2559-2567 (2003).
6. Z. Zhang, P. M. Harrison, Y. Liu and M. Gerstein. Millions of years of evolution preserved: a comprehensive catalog of the processed pseudogenes in the human genome. *Genome Res.* **13**(12), 2541-2558 (2003).
7. A. M. Weiner, P. L. Deininger and A. Efstratiadis. Nonviral retroposons: genes, pseudogenes, and transposable elements generated by the reverse flow of genetic information. *Annu Rev Biochem.* **55**, 631-661 (1986).
8. A. Pavlicek, J. Paces, R. Zika and J. Hejnar. Length distribution of long interspersed nucleotide elements (LINEs) and processed pseudogenes of human endogenous retroviruses: implications for retrotransposition and pseudogene detection. *Gene.* **300**(1-2), 189-194 (2002).
9. M. Hattori, S. Kuhara, O. Takenaka and Y. Sakaki. L1 family of repetitive DNA sequences in primates may be derived from a sequence encoding a reverse transcriptase-related protein. *Nature.* **321**(6070), 625-628 (1986).
10. D. D. Luan and T. H. Eickbush. RNA template requirements for target DNA-primed reverse transcription by the R2 retrotransposable element. *Mol Cell Biol.* **15**(7), 3882-3891 (1995).
11. G. J. Cost, Q. Feng, A. Jacquier and J. D. Boeke. Human L1 element target-primed reverse transcription in vitro. *EMBO J.* **21**(21),5899-5910 (2002).
12. D. Grover, M. Mukerji, P. Bhatnagar, K. Kannan and S. K. Brahmachari. Alu repeat analysis in the complete human genome: trends and variations with respect to genomic composition. *Bioinformatics.* **20**(6), 813-817 (2004).
13. International Human Genome Sequencing Consortium. Initial sequencing and analysis of the human genome. *Nature* **409,** 860-921 (2001).
14. Z. Zhang, N. Carriero and M. Gerstein. Comparative analysis of processed pseudogenes in the mouse and human genomes. *Trends Genet.* **20**(2), 62-67 (2004).
15. E. S. Balakirev and F. J. Ayala. Pseudogenes: are they "junk" or functional DNA? *Annu Rev Genet.* **37,** 123-151 (2003).
16. P. M. Harrison, D. Zheng, Z. Zhang, N. Carriero, M. Gerstein. Transcribed processed pseudogenes in the human genome: an intermediate form of expressed retrosequence lacking protein-coding ability. *Nucleic Acids Res.* **33**(8), 2374-2383 (2005).
17. J. R. McCarrey and K. Thomas. Human testis-specific PGK gene lacks introns and possesses characteristics of a processed gene. *Nature.* **326**(6112), 501-505 (1987).
18. J. Bristow, S. E. Gitelman, M. K. Tee, B. Staels and W. L. Miller. Abundant adrenal-specific transcription of the human P450c21A "pseudogene". *J Biol Chem.* **268**(17), 12919-12924 (1993).
19. S. Ramos-Onsins and M. Aguade. Molecular evolution of the Cecropin multigene family in Drosophila. functional genes vs. pseudogenes. *Genetics.* **150**(1), 157-171 (1998).
20. S. M. Troyanovsky and R. E. Leube. Activation of the silent human cytokeratin 17 pseudogene-promoter region by cryptic enhancer elements of the cytokeratin 17 gene. *Eur J Biochem.* **225**(1), 61-69 (1994).
21. O. Podlaha and J. Zhang. Nonneutral evolution of the transcribed pseudogene Makorin1-p1 in mice. *Mol Biol Evol.* **21**(12), 2202-2209 (2004).
22. B. S. Zhou, D. R. Beidler and Y. C. Cheng. Identification of antisense RNA transcripts from a human DNA topoisomerase I pseudogene. *Cancer Res.* **52**(15), 4280-4285 (1992).
23. D. Weil, M. A. Power, G. C. Webb and C. L. Li. Antisense transcription of a murine FGFR-3 psuedogene during fetal developement. *Gene.* **187**(1), 115-122 (1997).
24. I. E. Scheffler. Mitochondria make a come back. *Adv Drug Deliv Rev.* **49**(1-2), 3-26 (2001).
25. S. Berry. Endosymbiosis and the design of eukaryotic electron transport. *Biochim Biophys Acta.* **1606**(1-3), 57-72 (2003).
26. O. G. Berg and C. G. Kurland. Why mitochondrial genes are most often found in nuclei. *Mol Biol Evol.* **17**(6), 951-961 (2000).

27. C. G. Kurland and S. G. Andersson. Origin and evolution of the mitochondrial proteome. *Microbiol Mol Biol Rev.* **64**(4), 786-820 (2000).
28. S. F. Altschul, W. Gish, W. Miller, E. W. Myers and D. J. Lipman. Basic local alignment search tool. *J Mol Biol.* **215**(3), 403-410 (1990).
29. T. Gabaldon and M. A. Huynen. Shaping the mitochondrial proteome. *Biochim Biophys Acta.* **1659**(2-3), 212-220 (2004).
30. J. E. Walker. The NADH:ubiquinone oxidoreductase (complex I) of respiratory chains. *Q Rev Biophys.* **25**(3), 253-324 (1992).
31. U. Weidner, S. Geier, A. Ptock, T. Friedrich, H. Leif and H. Weiss. The gene locus of the proton-translocating NADH: ubiquinone oxidoreductase in Escherichia coli. Organization of the 14 genes and relationship between the derived proteins and subunits of mitochondrial complex I. *J Mol Biol.* **233**(1), 109-122 (1993).
32. J. Carroll, R. J. Shannon, I. M. Fearnley, J. E. Walker and J. Hirst. Definition of the nuclear encoded protein composition of bovine heart mitochondrial complex I. Identification of two new subunits. *J Biol Chem.* **277**(52), 50311-50317 (2002).
33. T. Gabaldon, D. Rainey and M. A. Huynen. Tracing the evolution of a large protein complex in the eukaryotes, NADH:ubiquinone oxidoreductase (Complex I). *J Mol Biol.* **348**(4), 857-870 (2005).
34. A. Khelifi, L. Duret and D. Mouchiroud. HOPPSIGEN: a database of human and mouse processed pseudogenes. *Nucleic Acids Res.* 33(Database issue), D59-D66 (2005).
35. K.A. Kreuzer, U. Lass, O. Landt, A. Nitsche, J. Laser, H. Ellerbrok, G. Pauli, D. Huhn and C.A Schmidt. Highly sensitive and specific fluorescence reverse transcription-PCR assay for the pseudogene-free detection of beta-actin transcripts as quantitative reference. *Clin Chem.* **45**(2), 297-300 (1999).
36. D. Zheng, Z. Zhang, P. M. Harrison, J. Karro, N. Carriero and M. Gerstein. Integrated pseudogene annotation for human chromosome 22: evidence for transcription. *J Mol Biol.* **349**(1), 27-45 (2005).

INTRATUMORAL VEGF AND FGF1 ADMINISTRATION ALTERS TUMOR GROWTH, VASCULAR DENSITY, OXYGENATION, AND EXPRESSION OF MCP-1 AND INTERLEUKINS

Paul Okunieff, Jianzhong Sun, Bruce Fenton, Weimin Liu, and Ivan Ding*

Abstract: The biological and physiological effects of exogenous FGF1 and VEGF were measured using the KHT murine fibrosarcoma tumor model. Tumor-bearing C3H mice were treated intratumorally with either one or six daily doses of 6 μg/mouse FGF1, VEGF, or saline. Tumors were excised 24 hrs after the final injection. Compared to controls, only FGF1 treatment significantly increased tumor weight and size, and only in the 6 dose group. Both FGF1 and VEGF administration (6 dose) decreased tumor cell hypoxia as detected by EF5 uptake: 85% ± 5% for FGF1 and 82% ± 6% for VEGF versus 100% ± 6% for controls. Decreased tumor cell EF5 staining, however, was not associated with changes in numbers of structural or angiogenic vessels. $DiOC_7$ staining showed a slight decrease in perfused vessel numbers in tumors treated with daily VEGF. Intratumoral injections of FGF1 or VEGF also slightly decreased the tumor tissue chemokine MCP-1, interleukins (IL-1β, IL-6, and IL-18) mRNA expression, and increased NFκB binding without altering Ap-1 binding of IκB protein expression. In summary, single pulse exposures of tumors to angiogenic factors had little or no effects on tumor growth or perfusion, while daily exposures stimulated tumor growth through improved tumor oxygenation. This improved vascular function occurs without an increase in vascular density.

1. INTRODUCTION

Investigators have shown that fibroblast growth factors (FGFs) and vascular endothelial growth factor (VEGF) have beneficial effects on normal tissues undergoing radiation treatment.[1-3] FGFs can radioprotect bone marrow,[1, 2] small bowel,[4] prevent bone growth defects,[5] and may reduce radiation-induced pneumonitis.[3, 6] VEGF has less dramatic effects on prevention of small bowel toxicity and hematopoietic toxicity. Protection is seen with systemic administration of cytokines at various times ranging from 24 hours before or after irradiation. Very little systemic cytokine is required to obtain maximum benefit.[1-3] These same cytokines, however, are frequently produced by tumors and

* Department of Radiation Oncology, University of Rochester Medical Center, Rochester, NY 14642 USA

contribute in varying degrees to tumor growth, invasion, and metastasis.[7-9] The role that these cytokines play is strongly related to dose, duration of exposure, and the balance between the various cytokines and their receptor expression levels.[10, 11] FGF1 and VEGF are also associated experimentally and clinically with tumor progression and metastasis.[8,9] Clinically, it is believed that these growth factors mediate the hypoxia and angiogenesis associated with tumor aggressiveness and metastatic spread,[12, 13] but most studies suggest that these cytokines are only deleterious when delivered chronically at high doses.[10, 11] Thus normal tissue radioprotection might be achieved at doses that would not be harmful. We examined tumor physiological alterations (vascular density, vascular perfusion, tumor proliferation, and oxygenation) induced by FGF1 and VEGF, given intratumorally either in a single dose or in six daily doses, in the murine KHT fibrosarcoma. Intratumoral injection was used to assure that very high levels of growth factors were delivered to the tumor. Each dose (6 μg/mouse/day) is sufficient to produce maximal radiation protection when delivered systemically.[1, 2, 4] Thus six daily treatments exceed the dose that might be required clinically for radioprotective therapy.

2. METHODS

2.1. Tumor Model

Single cell suspensions were obtained by mechanical disruption of KHT fibrosarcomas, and 5×10^5 cells were inoculated i.m. into the hind limbs of 6-8 week-old female C3H/Hej mice. Tumors were selected for treatment when they reached volumes of between 200 and 300 mm^3.

2.2. Angiogenic Agent Treatments

At appropriate tumor volumes, human recombinant FGF1 and VEGF (Pepro Tech Inc., Rocky Hill, NJ) was administrated intratumorally at 60 μg/ml in a volume of 100 μl (6μg/mouse/injection/day), and tumors were frozen one day after completion of treatment.

2.3. Measurement of Anatomical, Perfused, and Angiogenic Blood Vessels, Hypoxia, and Proliferation (PCNA)

To visualize blood vessels open to flow, an intravascularly injected stain, $DiOC_7$ was utilized.[14] CD31 staining was used for identification of total structural vessels. In addition, CD105 was used to identify neovessels. CD105 is strongly expressed in activated endothelial cells of various human tumors, including breast cancer, but weakly present in mature blood vessels of normal tissues.[15, 16] EF5 is a pentaflovinated hydrocarbon that undergoes irreversible covelent bonding to protein and DNA under hypoxic conditions. EF5 adducts can then be detected and quantified using fluorescent monoclonal antibodies and image analysis. EF5 was injected i.v. one hr before tumor freezing, and localized areas of tumor hypoxia were assessed by identification of EF5 metabolism,[17] using a fluorochrome (Cy3) conjugated antibody. Serial sections of each tumor were also stained with hematoxylin-eosin, and with anti-proliferating cell nuclear antigen (PCNA) antibody.[8] The numbers of PCNA-labeled cells were determined using image analysis techniques.

2.4. Imaging and Image Analysis

Frozen tumor sections were cut, stained, imaged, and analyzed using Image Pro software.[14] Each section was scanned under three staining conditions: $DiOC_7$ staining marked perfused vessels, CD31 marked anatomical blood vessels, and CD105 marked angiogenic vessels. Visual Basic macros were used to perform a "closest individual" analysis,[14] whereby the distances from computer-superimposed sampling points to the nearest blood vessel were determined.

2.5. Cytokine mRNA Determination by RNase Protection Assay

Total RNA was isolated by pulverizing the frozen tissue and dissolving it in TRIzol reagent. RNase protection was performed using established multi-probe template sets (Pharmingen, San Diego, CA). The MCK5 chemokine set includes: Lymphotactin, Rantes, Eotaxin, MIP-1α, MIP-1β, MIP-2, IP-10, MCP-1, and TCA-3. The MCK3 inflammatory cytokine set includes: tumor necrosis factor α and β, transforming growth factor β1,2,3, interferon β and γ, interleukin 6, and lymphotoxin-β. Two internal controls, L32 and GAPDH, were used. The quantitation of mRNA expression level for each sample was measured by phosphorimaging.

2.6. Nuclear Extract Preparation and Gel Shift Assay

Protein-DNA binding reactions were carried out for 20 min on ice, using 5 μg nuclear extract, 32P-labeled DNA probes for the NFκB consensus binding sequence, with or without 1 ng unlabeled competitor probe[18]. Complexes were separated by electrophoresis on nondenaturing 5% polyacrylamide gels and assayed by autoradiography and Phosphorimager analysis. Antibodies against p65 or p50 subunit of the NFκB complex were used for super-shift assays.

2.7. Western Blot Analysis

Cell lysates were prepared and separated by SDS-PAGE on 12% polyacrylamide gels and electrotransferred onto nitrocellulose membranes. Membranes were blocked in Tris-buffered saline containing 0.1% Tween 20 and 3% non-fat dry milk for 30 min at room temperature, then incubated overnight with the Ikb (1:1,000) antibody. Thereafter, gels were washed in Tris-buffered saline and incubated with horseradish peroxidase-conjugated goat anti-rabbit IgG in blocking solution for 1 h.

2.8. Statistical Methods

Median blood vessel distances and EF5/Cy3 intensities were compared using the unpaired Student's t-test or Mann-Whitney Rank Sum test (significant at $p<0.05$).

3. RESULTS

Two types of intratumoral exposure to FGF1 or VEGF were explored: an acute single injection or six daily intratumoral injections. As shown in Table 1, single exposures did not significantly alter tumor growth. Chronic exposure to FGF1, however, significantly enhanced both tumor size and tumor weight ($p<0.01$). The increase was not dramatic, with a size ratio of 1.36 ± 0.17 ($\pm$ SD) after six days of treatment. KHT tumors have a volume doubling time of ~1.5 days; thus the observed size ratio converts to a growth progression of, at most, a 10% faster volume doubling time. VEGF did not significantly increase tumor size, even with six daily treatments. Although PCNA labeling appeared slightly higher after six days (Figure 1b, 1d, 1f) the increase was not statistically significant. Histologically, tumor cell density also appeared to be higher in FGF1 (Figure 1c) and VEGF treated tumors (Figure 1e), compared to saline injected tumors (Figure 1a).

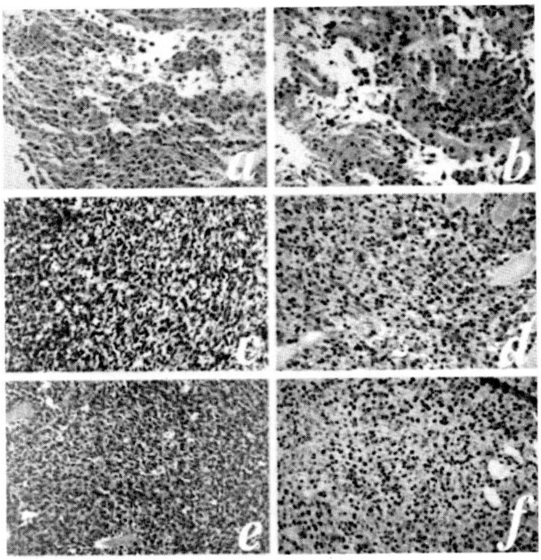

Figure 1. Tumor sections from H&E stained (a, c, e) and anti-PCNA antibody-stained slides (b, d, f). Mice were given 6 injections of FGF1 (c, d), VEGF (e, f) or saline (a, b). VEGF and FGF1 treated tumors were very cellular and PCNA labeling was abundant, but not significantly elevated compared with saline-treated tumors.

The effects of FGF1 and VEGF on tumor vasculature and oxygenation were determined immunohistochemically using EF5/Cy3, CD31, CD105, and $DiOC_7$ staining. EF5/Cy3 intensity in FGF1 or VEGF-treated tumor was normalized to fluorescence intensity of saline-treated tumors. Single injections had no significant effect on EF5/Cy3 staining intensity. However, tumors in the six injections group had significantly lower EF5/Cy3 staining intensity than the saline-treated tumors, suggesting reduced hypoxia in angiogenic growth factor-treated tumors (Table 1).

Structural and angiogenic vessels were measured by CD31 and CD105 staining, respectively, and functional vessels were measured by $DiOC_7$. No significant alterations in total structural vessels were observed after FGF1 or VEGF treatment compared to control (Table 1). The median distances to the nearest vessel were similar, and independent of growth factor or the number of injections performed. Vessel perfusion, however, was reduced by the injection process, and multiple injections led to a greater distance to the nearest perfused vessel in saline and cytokine treated animals. Although response to the trauma of the six injections themselves might explain the reduction in perfusion measured by $DiOC_7$, in the single injection group there was no evidence that perfusion was enhanced by angiogenic factors.

Table 1. Pathophysiological and biological parameters in tumor-bearing mice following intratumoral saline, FGF1, or VEGF treatment.

	Saline	FGF1	VEGF
	One Injection (n=3-5)		
Tumor size (mm^3)	568±49	700±37	588±39
CD31-vessel spacing	242±1.1	27.0±1.6	25.7±1.5
CD105-vessel spacing	34.5±3.1	39.8±4.0	45.2±4.3
Normalized EF5/Cy3 intensity	1.00±0.05	0.87±0.05	0.92±0.06
	Six Injections (n=3-5)		
Tumor size (mm^3)	439±37	596±56*	495±57
Tumor weight (g)	0.69±0.09	0.93±0.02**	0.87±0.01
PCNA-labeled nuclei/mm^2	2612±930	3410±86	3899±288
CD31-vessel spacing	29.0±1.6	26.4±1.5	26.1±1.1
CD105-vessel spacing	31.8±1.6	31.1±2.2	26.4±0.9
Normalized EF5/Cy3 intensity	1.00±0.06	0.85±0.05*	0.82±0.06*

* $p<0.05$; ** $p<0.01$

Tumor tissue is also comprised of non-malignant stroma and inflammation. The stromal composition of tumor is controlled by chemotaxis of inflammatory cells, endothelial cells and fibroblasts, and the synthesis of matrix. These processes are governed, in turn, by chemokines and cytokines that regulate chemotaxis and inflammation. Gene expression for inflammatory and chemotactic cytokines was measured using RNase protection. Following six doses of saline, MCP-1 declined to 50% and 73% of control for VEGF and FGF1-treated tumors respectively (Figure 2B). The decreased MCP-1 was associated

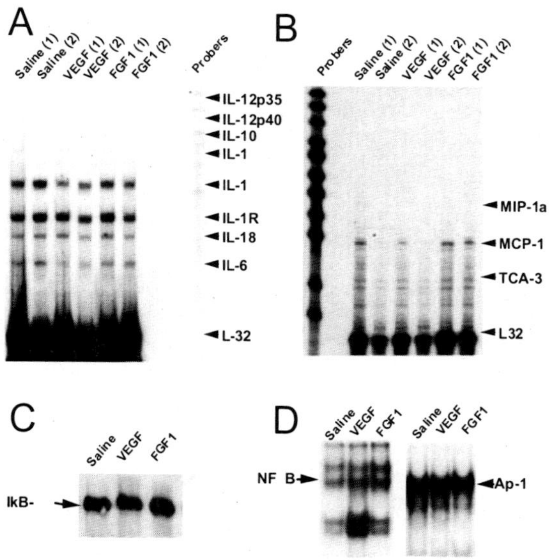

Figure 2: RNA extracts of two tumors from each treatment group were analyzed by RNase protection. Interleukin-1β, IL-6 and IL-18 (A) and MCP-1 (B) were decreased in VEGF-treated tumors. Intratumor cytokine treatments were not associated with a measurable change of Ap1 binding (D) or IκB-α expression (C), although there was an elevation of NFkB binding (D).

with a decrease in IL-1β (40% for FGF1 and 51% for VEGF), IL-6 (30% for FGF1 and 44% for VEGF), and IL-18 (59% for FGF1 and 66% for VEGF). Ap-1 binding (Figure 2D) and IκB-α protein expression (Figure 2C) were unaffected by injection of FGF1 and VEGF. NFκB binding, however, was increased in the FGF1 and VEGF-treated tumors (Figure 2D).

4. DISCUSSION

Previous studies have suggested that acute exposures of tumor to growth factors may have no adverse effects, and that chronic exposures have little added normal tissue advantage compared to acute exposure. Disadvantageous effects of angiogenic growth factors have been reported by several investigator.[8, 13, 22, 23] Leith et al. found that intraperitoneal injection of FGF1 (0.25mg/kg X 7 days), in nude mice bearing human colon xenografts, dramatically increased tumor volume and decreased tumor hypoxia.[24, 25] Enhancement of other types of tumor growth by other growth factors including epidermal growth factor (EGF), has also been reported.[26-28] It was suggested that decreased tumor hypoxia after growth factor treatment was due to improved tumor perfusion. In our study, we demonstrated an increased KHT tumor weight and volume after six daily FGF1 doses, and slightly elevated PCNA labeling. The increased tumor growth was associated with decreased tumor hypoxia after FGF1 or VEGF, consistent with the previous findings. The decreased hypoxia, however, was not correlated with increased total structural, angiogenic or perfused vessels. This apparently contradictory finding can be explained by: 1) a reduction of metabolic rate in the angiogenic factor-treated tumors leading to reduced oxygen consumption, 2) FGF or VEGF-mediated tumor interstitial space expansion with resultant decreased cellularity, or 3) a higher flow rate within the functioning vessels. We favor the third explanation. Metabolic rate of tumor is expected to be at least as high considering the high PCNA labeling, and the pathological observations do not indicate increased extracellular volume in angiogenic growth factor-treated tumors. Hence, the first two explanations are unlikely. High levels of VEGF or FGF1, however, could reasonably result in improved functioning of existing vasculature.

Tumors with high levels of angiogenic growth factors appear to be more invasive and metastatic,[8, 9] as confirmed by counts of vascular hot-spots.[29] Tumor cells around vascular hot-spots commonly express high levels of FGFs or VEGF. CD31 and CD105 staining, however, did not demonstrate increased vascularity after single or multiple injections of FGF1 or VEGF in KHT tumors. Recent clinical studies also suggest that high systemic levels of FGF in blood or urine lead to more metastatic bladder[30] and other malignancies.[31] Although enlarged tumor size was seen after FGF1, tumors were not significantly larger after VEGF. This suggests that: 1) baseline circulating levels of VEGF are already elevated, or 2) tumors may already be maximally stimulated to produce high levels of several growth factors. Under these conditions, the KHT vasculature may be unable to respond to acute exposure of another growth factor.

Tumor hypoxia is considered an important determinant of metastatic disease for carcinoma of the uterine cervix,[32] and recent studies have shown that patients with soft tissue sarcoma are at much higher risk of developing metastatic disease if their tumors are hypoxic.[33] Although the mechanism is uncertain, it is known that hypoxia induces VEGF and FGF, and that these elevated angiogenic growth factors can in turn promote aggressive tumor phenotypes. Multiple injections of FGF1 or VEGF decreased EF5 staining in the KHT sarcoma model, indicating an increase in tumor oxygenation. But, tumors with high levels of VEGF or FGF1 are expected to have lower levels of hypoxia than those with low VEGF or FGF1 expression. This paradox can be explained if we assume that hypoxia chronically induces tumor cells to produce VEGF and FGF, whereas substantial acute injections of the growth factors may be able to reduce hypoxia.

Intratumoral injection of FGF1 or VEGF decreased mRNA expression of the MCP-1 chemokine and IL-1β, IL-6, and IL-18 inflammatory cytokines, although the cause remains unknown. MCP-1 suppression, however, is expected to result in reduced chemokine/cytokine production and less activation of infiltrating mononuclear cells, which in turn cause reduced activation of inflammatory cytokines. This reduced inflammation could better preserve vascular function. Intracellular signaling molecules involved in cytokine signaling were also measured: NFκB binding was found to be elevated after growth factor treatment, while IκB protein was unaffected.

Finally, we attempted to ensure that high levels of exogenous growth factor reached the tumor through direct tumor injection. The daily delivery of 100 μl into the tumor, whether saline or growth factor, was clearly traumatic and could have blunted some of the growth and perfusion responses. Thus it is premature to conclude that short courses of growth factors can be safely administered to cancer patients to protect normal tissues specifically.

In summary, multiple doses of VEGF had minimal or no effects on tumor proliferation, but increased endothelial proliferation. This resulted in increased angiogenesis, improved oxygenation, and a possible decrease in chemokine and inflammatory cytokine response. In contrast, multiple doses of intratumoral FGF1 most likely had proliferative effects on both KHT tumor cells and endothelium, resulting in stable vascular density. Oxygenation was increased, and changes in chemokines and cytokines, if any, were similar to those following VEGF. Single doses of either VEGF or FGF1 did not have significant effects on tumor or vascular proliferation. These data suggest that if angiogenic factors are given clinically to reduce normal tissue consequences of therapy, care must be taken to avoid tumor growth, and such factors should be given in single doses or for very short courses.

5. REFERENCES

1. I. Ding, K. Huang, X. Wang, J. R. Greig, R. W. Miller, and P. Okunieff P, Radioprotection of hematopoietic tissue by fibroblast growth factors in fractionated radiation experiments, Acta. Oncol.36, 337-340 (1997).
2. I. Ding, T. Wu, H. Matsubara, et al, Acidic fibroblast growth factor (FGF1) increases survival and haematopoietic recovery in total body irradiated C3H/HeNCr mice, Cytokine 9, 59-65 (1997).
3. Z. Fuks, R. S. Persaud, A. Alfieri, et al, Basic fibroblast growth factor protects endothelial cells against radiation-induced programmed cell death in vitro and in vivo, Cancer Res. 54, 2582-2590 (1994).
4. P. Okunieff, M. Mester, J. Wang, et al, In vivo radioprotective effects of angiogenic growth factors on the small bowel of C3H mice, Radiat. Res. 150, 204-211 (1998).
5. X. Wang, I. Ding, H. Xie H, et al, Hyperbaric oxygen and basic fibroblast growth factor promote growth of irradiated bone, Int. J. Radiat. Oncol. Biol. Phys. 40, 189-196 (1998).
6. P. G. Tee , E. L. Travis, Basic fibroblast growth factor does not protect against classical radiation pneumonitis in two strains of mice, Cancer Res. 55, 298-302 (1995).
7. M. Presta, D. B. Rifkin, New aspects of blood vessel growth: tumor and tissue-derived angiogenesis factors, Haemostasis 18, 6-17 (1988).
8. L. Zhang, S. Kharbanda, D. Chen, et al, MCF-7 breast carcinoma cells overexpressing FGF-1 form vascularized, metastatic tumors in ovariectomized or tamoxifen-treated nude mice, Oncogene 15, 2093-2108 (1997).
9. S. W. McLeskey, J. Kurebayashi, S. F. Honig, et al, Fibroblast growth factor 4 transfection of MCF-7 cells produces cell lines that are tumorigenic and metastatic in ovariectomized or tamoxifen-treated athymic nude mice, Cancer Res. 53, 2168-2177 (1993).

10. R. Neta, Modulation of radiation damage by cytokines, Stem Cells 15 (Suppl 2), 87-94 (1997).
11. R. Neta, Modulation with cytokines of radiation injury: suggested mechanisms of action, Environ. Health Perspect. 105(Suppl 6), 1463-1465 (1997).
12. J. Folkman, Angiogenesis: initiation and control, Ann. N. Y. Acad. Sci. 401, 212-227 (1982).
13. F. G. Kern, M. E. Lippman, S. W. McLeskey, et al, The role of angiogenic growth factors in breast cancer progression, Cancer Metastas. Rev. 15, 213-219 (1996).
14. B. M. Fenton, S. F. Paoni, J. Lee, C. J. Koch, E. M. Lord, Quantification of tumour vasculature and hypoxia by immunohistochemical staining and HbO_2 saturation measurements, Br. J. Cancer 79, 464-471 (1999).
15. B. Bodey, B. Bodey Jr., S. E. Siegel, H. E. Kaiser, Immunocytochemical detection of endoglin is indicative of angiogenesis in malignant melanoma, Anticancer Res. 18, 2701-2710 (1998).
16. B. Bodey, B. Bodey Jr., S. E. Siegel, H. E. Kaiser, Over-expression of endoglin (CD105): a marker of breast carcinoma- induced neo-vascularization, Anticancer Res. 18, 3621-3628 (1998).
17. E. M. Lord, L. Harwell, C. J. Koch, Detection of hypoxic cells by monoclonal antibody recognizing 2-nitroimidazole adducts, Cancer Res. 53, 5721-5726 (1993).
18. A. M. Manning, F. P. Bell, C. L. Rosenbloom, et al, NF-kappa B is activated during acute inflammation in vivo in association with elevated endothelial cell adhesion molecule gene expression and leukocyte recruitment, J. Inflamm. 45, 283-296 (1995).
19. G. Molema, D. K. Meijer, L. F. de Leij, Tumor vasculature targeted therapies: getting the players organized, Biochem. Pharmacol. 55, 1939-1945 (1998).
20. P. Okunieff, X. Wang, M. Li M, et al, Chronic radiation bone toxicity is associated with decreased perfusion without elevation of circulating or soft tissue TGF beta or TNF alpha, Adv. Exp. Med. Biol. 454, 325-333 (1998).
21. P. Okunieff, E. H. Abraham, M. Moini, et al, Basic fibroblast growth factor radioprotects bone marrow and not RIF1 tumor, Acta. Oncol. 34, 435-438 (1995).
22. F. G. Kern, S. W. McLeskey, L. Zhang, et al, Transfected MCF-7 cells as a model for breast-cancer progression, Breast Cancer Res. Treat. 31, 153-165 (1994).
23. J. Yoneda, H. Kuniyasu, M. A. Crispens, J. E. Price, C. D. Bucana, I. J. Fidler, Expression of angiogenesis-related genes and progression of human ovarian carcinomas in nude mice, J. Natl. Cancer Inst. 90, 447-454 (1998).
24. J. T. Leith, S. Michelson, Effects of administration of basic fibroblast growth factor on hypoxic fractions in xenografted DLD-2 human tumours: time dependence, Br. J. Cancer 68, 727-731 (1993).
25. J. T. Leith, G. Papa, L. Quaranto, S. Michelson, Modification of the volumetric growth responses and steady-state hypoxic fractions of xenografted DLD-2 human colon carcinomas by administration of basic fibroblast growth factor or suramin, Br. J. Cancer 66, 345-348 (1992).
26. J. T. Leith, G. Padfield, S. Michelson, Effects of partial hepatectomy on the growth characteristics and hypoxic fractions of xenografted DLD-2 human colon cancers, Radiat. Res. 132, 263-268 (1992).
27. J. T. Leith, P. Harrigan, G. Padfield, L. Faulkner, S. Michelson, Modification of the growth rates and hypoxic fractions of xenografted A431 tumors by sialoadenectomy or exogenously supplied epidermal growth factor, Cancer Res. 51, 4111-4113 (1991).
28. J. T. Leith, Modification of the hypoxic fraction of a xenografted human colon tumor by differentiation-inducing agents, J. Natl. Cancer Inst. 80, 444-447 (1988).
29. A. Balsari, J. A. Maier, M. I. Colnaghi, S. Menard, Correlation between tumor vascularity, vascular endothelial growth factor production by tumor cells, serum vascular endothelial growth factor levels, and serum angiogenic activity in patients with breast carcinoma, Lab. Invest. 79, 897-902 (1999).
30. D. Strohmeyer, Pathophysiology of tumor angiogenesis and its relevance in renal cell cancer, Anticancer Res. 19, 1557-1561 (1999).
31. F. Kaplan, J. Sawyer, S. Connors, K. Keough, E. Shore, F. Gannon, D. Glaser, D. Rocke, M. Zasloff, J. Folkman, Urinary basic fibroblast growth factor. A biochemical marker for preosseous fibroproliferative lesions in patients with fibrodysplasia ossificans progressiva, Clin. Orthop. 346, 59-65 (1998).
32. P. Vaupel, D. K. Kelleher, M. Höckel, Oygen status of malignant tumors: pathogenesis of hypoxia and significance for tumor therapy, Semin. Oncol. 28(2 Suppl 8), 29-35 (2001).
33. D. M. Brizel, S. P. Scully, J. M. Harrelson, L. J. Layfield, J. M. Bean, L. R. Prosnitz, M. W. Dewhirst, Tumor oxygenation predicts for the likelihood of distant metastases in human soft tissue sarcoma, Cancer Res. 56(5), 941-943 (1996).

NITRIC OXIDE IN THE KIDNEY

Direct measurements of bioavailable renal nitric oxide

Fredrik Palm, Lina Nordquist, and Donald G. Buerk*

Abstract: Increasing efforts have been directed towards investigating the involvement of nitric oxide (NO) for normal kidney function. Recently, a crucial role of NO in the development of progressive renal dysfunction has been reported during diabetes and hypertension.

Indirect estimation of renal NO production include urinary nitrite/nitrate measurements, but there are several disadvantages of indirect methods since production and bioavailability of NO rarely coincide. Thus, direct measurement of *in vivo* NO bioavailability is preferred, although these methods are more time consuming and require highly specialized equipment and knowledge.

This review focuses on two techniques for *in vivo* measurement of bioavailable NO in the kidney. We have applied Whalen-type recessed NO microsensors for measurement of NO in the kidney cortex, whereas the hemoglobin-trapping technique seems to be more suitable for NO measurement in the renal medulla. Both methods are robust and reliable, and we discuss advantages and shortcomings of each method.

1. INTRODUCTION

Nitric oxide is a free radical that mediates cell signaling, host defense and inflammatory functions, as well as regulating hemodynamics and energetics in almost all cells studied so far. The role of nitric oxide (NO) in the regulation of vascular tone and cellular metabolism has gained increasing attention during the last few decades. In the kidney, NO also regulates kidney function and has thus been shown to play a pivotal role in the

* Fredrik Palm, Lina Nordquist, Department of Medical Cell Biology, Uppsala University, Biomedical Center, PO 571, SE 751 23 Uppsala, Sweden.
Donald G. Buerk, Departments of Physiology and Bioengineering, University of Pennsylvania School of Medicine, Philadelphia, Pennsylvania, USA

long-term regulation of systemic blood pressure[1]. Renal NO is derived from all three known nitric oxide synthase (NOS) isoforms[1], although it has also been reported that non-enzymatic sources can contribute substantially during certain conditions[2].

Evidence for the important role of NO in maintaining normal kidney function is mainly derived from studies where the renal NO system has been altered. This often results in altered kidney function, with secondary systemic effects. For example, the renal NO system is altered during hypertension[3], diabetes[4] and renal ischemia-reperfusion[2].

The kidney is highly heterogeneous, both with regard to the anatomical structures and the internal milieu, resulting in several challenges when measuring local renal NO concentration. Furthermore, the short half life of NO and the low physiological concentrations provide further challenges, in addition to the differences in regulation of the NO system in the different parts of the kidney. Several techniques have been developed and evaluated in order to measure local NO concentration, but only a few of these are suitable for directly detecting renal NO concentrations *in vivo*.

2. PHYSIOLOGY AND REGULATION OF NITRIC OXIDE IN THE KIDNEY

All three NOS isoforms are present in the kidney, although in different anatomical structures within the renal architecture (Table 1)[1]. The spatial separation of the different NOS isoforms reflects different function of NO depending on the site of production and release. The reported NO concentrations are in the renal cortex in the range 35-90 nmol/l and in the renal medulla 55-130 nmol/l[5-7]. Thus, the special distribution of bioavailable NO *in vivo* is believed to be highly heterogeneous within the kidney, although data supporting this is still rudimental. The tubular reabsorption of electrolytes has been found to be tightly regulated by NO[8]. Reduced NO function in the renal medulla, which affects the medullary reabsorption of electrolytes, contributes to the salt sensitivity and development of hypertension in Dahl salt sensitive rats[3]. Furthermore, NO derived from the nNOS located in the macula densa alters the effectiveness of proximal reabsorption and is involved in the diabetes-induced glomerular hyperfiltration[9, 10].

Table 1. Distribution of the NOS isoforms in the kidney during normal physiological conditions[a].

NOS isoform	Normal distribution in the kidney
neuronal NOS (nNOS)	Macula densa Collecting ducts Renal pelvis
inducible NOS (iNOS)	Proximal tubules
endothelial NOS (eNOS)	Endothelium of afferent and efferent arterioles, and vasa recta Proximal tubules Thick ascending limbs Collecting ducts

[a] Modified from Kone and Baylis[1].

2.1. NO and Inhibition of Oxygen Consumption

Koivisto and co-workers have shown that NO inhibits the mitochondrial respiration of the proximal tubular cells, and that the inhibition is proportional to the inverse square of the oxygen concentration[11]. During normal physiological conditions, the influence of NO inhibition is likely to be significant in the renal medulla due to the normally low tissue oxygen tension (pO_2). We have recently demonstrated that bioavailability of NO in streptozotocin-diabetic animals is markedly lower compared to non-diabetic animals[5]. The reason for this decreased NO activity was identified as reduced substrate availability, which could be compensated for by administering excess L-arginine. Normalizing renal NO concentration with L-arginine also markedly increased renal tissue pO_2, probably due to the inhibition of mitochondrial oxygen consumption by NO.

2.2. Regulation of NOS Activity by Endogenous Inhibitors

Asymmetric dimethylarginine (ADMA) impairs NO either by competing with the NOS substrate L-arginine for membrane transport cationic aminoacid transporters, y^+, or by competitive inhibition of NOS. Therefore, increased ADMA concentration is proposed as a potential mechanism for the defective NO responses contributing to cardiovascular disease[12]. ADMA is metabolized by dimethylarginine dimethyl-aminohydrolases (DDAH-1 and -2)[13], where DDAH-1 is heavily expressed in the vascular endothelium and proximal tubules, and DDAH-2 in the macula densa cells and distal nephron[13]. Moreover, altered DDAH activity has been reported during pathologic conditions[14]. Inhibition of the NOS by ADMA can be overcome by excess administration of L-arginine, since the NOS inhibition is competitive.

3. MEASUREMENTS OF RENAL BIOAVAILABLE NITRIC OXIDE

3.1. Recessed Whalen-type NO Microsensors

Renal cortical NO concentrations can easily be detected using Nafion-coated Whalen-type NO microsensors first described by Buerk *et al*[15]. These microsensors are constructed in a manner similar to the recessed oxygen microelectrodes first described by Whalen and co-workers[16]. Glass micropipettes, pulled to a tip diameter less than 10 μm, are filled with Woods metal alloy (melting point about 78° C). Approximately 100 μm of the tip is left empty to allow gold to be plated inside the recess. The tip is then ground at an angle of 45 degrees to facilitate penetration into tissue. The tip is immersed in Nafion® perfluorinated resin (5% wt dissolved in aliphatic alcohols; Sigma-Aldrich) and allowed to dry in air. The membrane formed inside the recess is thus protected from physical damage, and will prevent interference from charged molecules. The electrodes are polarized at a constant voltage of +700 to +850 mV relative to a Ag/AgCl-reference. The electrochemical oxidation reaction for this electrode is:

$$NO + 4OH^- \rightarrow NO_3^- + 2H_2O + 3e^-$$

The oxidation current was amplified using Keithley 600-series Solid State Electrometers (Keithley Instruments, Cleveland, Ohio). These instruments can suppress high background currents, while still maintaining high resolution. Calibrations for each microsensor are done *in vitro* at controlled temperature using deoxygenated buffer bubbled with either nitrogen or a known NO concentration (Fig. 1)[15].

Only relative values of bioavailable NO concentration can be obtained, due to the unknown influence of interfering substances *in vivo*. However, by administrating high doses of a nonselective NOS inhibitor, e.g. Nω-nitro-L-arginine-methyl-ester (L-NAME) the decrease in NO concentration can be used to estimate baseline concentrations[5].

It should be noted that relatively high NO concentrations, especially in the renal medulla, can also originate from non-enzymatic sources and should be accounted for[2].

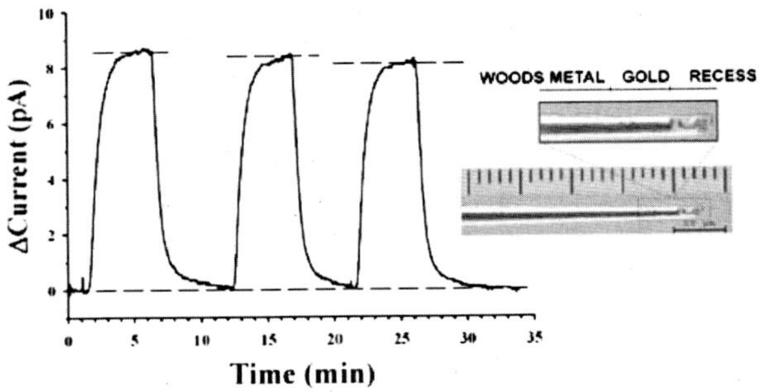

Figure 1. Left: Typical calibration curve for a recessed Whalen-type nitric oxide microsensor. Right: Recessed nitric oxide electrode. The magnified tip shows the recess and the gold plated on top of the Woods metal.

Substances influencing *in vivo* measurements of NO are preferentially ascorbic acid, pH, catecholamines and high concentrations of electrolytes, but due to the recess in the tip of the electrode and the use of the highly selective Nafion® membrane, this influence will be kept at an absolute minimum[15]. Furthermore, alterations in the NO concentrations are much faster than any physiological process responsible for changes in any of the interfering substances. The microsensors are influenced by temperature. However, this is only a potential problem if the temperature changes during the measurements.

The current sensitivity is approximately 350 nmol/picoampere for the calibration in Fig. 1, and depends on relative dimensions of the tip and recess. Detection limits can be below 1 nmol/l.

3.2. The Hemoglobin-trapping Technique

Renal medullary NO concentrations can be measured using a hemoglobin-trapping technique, which utilizes that NO stoichiometrically interacts with oxyhemoglobin (Oxy-Hb) forming methemoglobin (Met-Hb)[6, 17, 18]. The technique has been shown in several investigations to be the method of choice for NO measurements in the renal medulla[6], since this technique is not influenced by the high renal medullary osmolarity.

Oxy-Hb (3.0 μmol/l; Human A0 hemoglobin (ferrus); Sigma-Aldrich) is dissolved in Krebs buffer, modified according to the conditions in which the microdialysis probe will be placed. Typically, the most important parameters are pH and osmolarity, which in the renal medulla range from pH 7.35 to 7.20[19] and osmolarity of 600 to 1200 mOsm/kg H_2O. A Krebs buffer containing (in mmol/l) KCl 5.0, KH_2PO_4 1.7, $CaCl_2$ 2.5, $NaHCO_3$ 2.6, osmolarity adjusted to 800 mOsm/lkg by addition of NaCl, and pH adjusted to 7.35 is suitable for measurements of renal outer medullary NO concentrations[17, 19]. Fresh Oxy-Hb solution should be prepared daily and kept in a temperature controlled (0°C), gastight system in order to minimize autooxidation. We have also found that the Oxy-Hb will autooxidize when coming in contact with metal surfaces. This can be avoided by using only plastic components. A highly suitable fully plastic microdialysis probe is the Agnthos PES membrane cut off 6.000 Da, (Agnthos, Stockholm, Sweden). These probes can be ordered at custom membrane length. The 1 mm long with the 0.5 mm diameter is suitable for local renal medullary measurements and when perfused at a rate of 2.0 μl/min the detection limit is below 10 nmol/l of NO. This is well below normal concentrations found in the renal tissue[5, 6].

Although being very accurate and sensitive with a detection limit below 10 nmol/l, the hemoglobin-trapping technique needs about 30 minutes of continuous perfusion of the microdialysis probe since the smallest sample possible to measure in a spectrophotometer is typically in the range of 50 μl. Furthermore, after inserting the microdialysis probe into the renal tissue, the system needs a minimum of 60 minutes for stabilization. The entire experimental setup should also be shielded from any direct light source, since this can cause autooxidation of the Oxy-Hb and thus interfere with the measurement[17].

The absorbance maximum (λ_{max}) for Oxy-Hb is 415 nm and for Met-Hb 406 nm[17]. As previously described, the influence of spontaneous hemoglobin degradation can be corrected by using the absorbance at the isosbestic point 411 nm as reference zero[17]. To compensate for spectral distortion, the NO concentration is calculated by integrating the areas that represent the net increase in Met-Hb (absorbance at 394 nm to 411 nm) and the net decrease in Oxy-Hb (absorbance at 411 nm to 430 nm) spectrophotometrically[17]. The typical influence of autooxidation in this system is less than 5% of the total signal at NO concentrations of 10 mmol/l. Calibration is performed using several known concentrations of NO in deoxygenized Krebs buffer, typically within the expected physiological range (suggested calibrations: 10, 25, 50, 100, 200 nmol/l). Stable NO concentrations for sustained periods of time can be achieved by using a gastight deoxygenized system, which allows consecutive injections of aqueous NO solutions using a Hamilton syringe. The stock NO solution is prepared by bubbling deoxygenized Krebs buffer with NO gas of a known concentration.

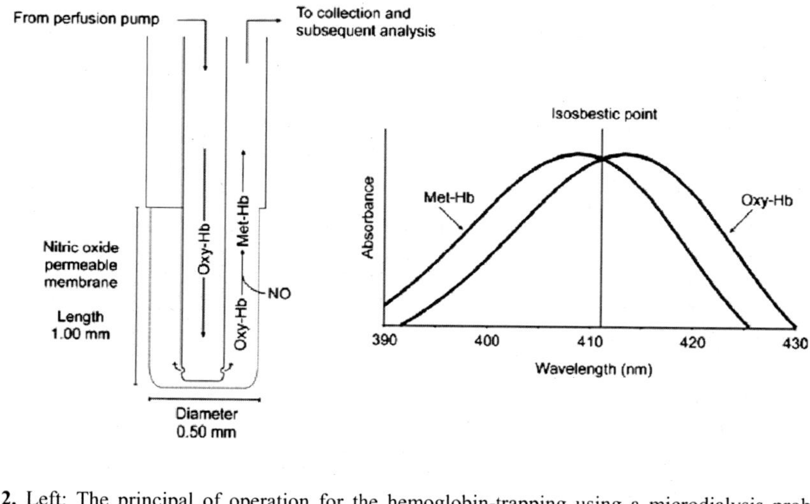

Figure 2. Left: The principal of operation for the hemoglobin-trapping using a microdialysis probe. Right: Typical absorbance patterns of oxy-hemoglobin (Oxy-Hb) and met-hemoglobin (Met-Hb).

3.3. Bare Carbon Fibers

Carbon fibers with diameters of approximately 30 μm have previously been used to measure renal NO concentrations[2, 20]. The theory of operation by electrochemical oxidation is identical to the recessed Whalen-type microsensors described above. However, due to the fragile properties of the carbon fiber *per se*, and the exposed membrane surrounding the fiber, these carbon fibers possess no advantage compared to the Whalen-type microsensors. Furthermore, the integrity of the Nafion® layer has to be verified after each experiment in order to ensure true recordings.

4. SUMMARY

The interest in renal NO is steadily growing and increasing efforts are devoted to direct NO measurements in the kidney. The rapid scavenging of NO, combined with the normally low physiological concentrations found, provide several challenges when choosing the method for *in vivo* measurements. Furthermore, the heterogenic renal internal milieus implicate that different detection techniques should be considered depending on where the NO is to be measured. According to our experience of detecting NO concentrations in the renal cortex, the method of choice is recessed Whalen-type microsensors. These have a superior time resolution and can be manufactured with a tip diameter of only a few μm. Due to the inherited difficulties of amperometric detection of NO in a fluctuating hyperosmotic milieu, the method of choice for renal medullary NO measurements seems to be the hemoglobin-trapping technique utilizing microdialysis.

5. ACKNOWLEDGEMENTS

Fredrik Palm is funded by the Swedish Medical Research Council, the Swedish Diabetes Association, and the Swedish Society for Medical Research, Lina Nordquist by the Swedish Society for Renal Medicine, and the Pharmaceutical Society, and Donald G. Buerk by the National Institutes of Health (Grant R01 HL-068164).

6. REFERENCES

1. B. C. Kone, and C. Baylis, Biosynthesis and homeostatic roles of nitric oxide in the normal kidney, *Am J Physiol* **272**(F561-F578 (1997).
2. M. G. Salom, B. Arregui, L. F. Carbonell, F. Ruiz, J. L. Gonzalez-Mora, and F. J. Fenoy, Renal ischemia induces an increase in nitric oxide levels from tissue stores, *Am J Physiol Regul Integr Comp Physiol* **289**(5), R1459-1466 (2005).
3. N. Miyata, A. P. Zou, D. L. Mattson, and A. W. Cowley, Jr., Renal medullary interstitial infusion of L-arginine prevents hypertension in Dahl salt-sensitive rats, *Am J Physiol* **275**(5 Pt 2), R1667-1673 (1998).
4. R. Komers, and S. Anderson, Paradoxes of nitric oxide in the diabetic kidney, *Am J Physiol Renal Physiol* **284**(6), F1121-F1137 (2003).
5. F. Palm, D. G. Buerk, P. O. Carlsson, P. Hansell, and P. Liss, Reduced nitric oxide concentration in the renal cortex of streptozotocin-induced diabetic rats: effects on renal oxygenation and microcirculation, *Diabetes* **54**(11), 3282-3287 (2005).
6. A. P. Zou, and A. W. Cowley, Jr., Nitric oxide in renal cortex and medulla. An in vivo microdialysis study, *Hypertension* **29**(1 Pt 2), 194-198 (1997).

7. M. Kakoki, H. S. Kim, W. J. Arendshorst, and D. L. Mattson, L-Arginine uptake affects nitric oxide production and blood flow in the renal medulla, *Am J Physiol Regul Integr Comp Physiol* **287**(6), R1478-1485 (2004).
8. P. A. Ortiz, and J. L. Garvin, Role of nitric oxide in the regulation of nephron transport, *Am J Physiol Renal Physiol* **282**(5), F777-84 (2002).
9. A. Deng, C. M. Miracle, J. M. Suarez, M. Lortie, J. Satriano, S. C. Thomson, K. A. Munger, and R. C. Blantz, Oxygen consumption in the kidney: Effects of nitric oxide synthase isoforms and angiotensin II, *Kidney Int* **68**(2), 723-730 (2005).
10. C. G. Schnackenberg, Physiological and pathophysiological roles of oxygen radicals in the renal microvasculature, *Am J Physiol Regul Integr Comp Physiol* **282**(2), R335-R342 (2002).
11. A. Koivisto, J. Pittner, M. Froelich, and A. E. Persson, Oxygen-dependent inhibition of respiration in isolated renal tubules by nitric oxide, *Kidney Int* **55**(6), 2368-2375 (1999).
12. R. H. Boger, Asymmetric dimethylarginine (ADMA) modulates endothelial function--therapeutic implications, *Vasc Med* **8**(3), 149-51 (2003).
13. C. T. Tran, J. M. Leiper, and P. Vallance, The DDAH/ADMA/NOS pathway, *Atheroscler Suppl* **4**(4), 33-40 (2003).
14. R. H. Boger, Asymmetric dimethylarginine, an endogenous inhibitor of nitric oxide synthase, explains the "L-arginine paradox" and acts as a novel cardiovascular risk factor, *J Nutr* **134**(10 Suppl), 2842S-2847S; discussion 2853S (2004).
15. D. G. Buerk, C. E. Riva, and S. D. Cranstoun, Nitric oxide has a vasodilatory role in cat optic nerve head during flicker stimuli, *Microvasc Res* **52**(1), 13-26 (1996).
16. W. J. Whalen, J. Riley, and P. Nair, A microelectrode for measuring intracellular PO_2, *J Appl Physiol* **23**(5), 798-801 (1967).
17. Y. Zhang, F. E. Samson, S. R. Nelson, and T. L. Pazdernik, Nitric oxide detection with intracerebral microdialysis: important considerations in the application of the hemoglobin-trapping technique, *J Neurosci Methods* **68**(2), 165-173 (1996).
18. A. Balcioglu, and T. J. Maher, The measurement of nitric oxide release induced by kainic acid using a novel hemoglobin-trapping technique with microdialysis, *Ann N Y Acad Sci* **738**(282-288 (1994).
19. F. Palm, P. Hansell, G. Ronquist, A. Waldenstrom, P. Liss, and P. O. Carlsson, Polyol-pathway-dependent disturbances in renal medullary metabolism in experimental insulin-deficient diabetes mellitus in rats, *Diabetologia* **47**(7), 1223-1231 (2004).
20. C. Thorup, M. Kornfeld, J. M. Winaver, M. S. Goligorsky, and L. C. Moore, Angiotensin-II stimulates nitric oxide release in isolated perfused renal resistance arteries, *Pflugers Arch* **435**(3), 432-434 (1998).

SEPARATION OF PROTEIN C FROM COHN FRACTION IV-1 BY MINI-ANTIBODY

Samin Rezania, Doh G. Ahn and Kyung A. Kang*

Abstract: Human protein C (PC) is a natural anticoagulant, antithrombotic, anti-inflammatory, and anti-apoptotic in the bloodstream. PC deficiency can lead to abnormal blood clot formation inside blood vessels, possibly causing heart attack, stroke, skin necrosis, or even death. PC can be, therefore, a valuable therapeutic with little side effect, unlike the currently used anti-coagulants. To reduce the cost involved in immuno purification of PC from blood plasma, single chain variable fragments (mini-Mab) are being produced by recombinant *E. coli* using phagemid technique. As an economic means of purifying the PC specific mini-Mab, metal affinity chromatography (IMAC) purification process was also investigated. Then using the purified mini-Mab, the feasibility of PC purification from the Cohn Fraction IV-1 was examined. Cohn Fraction IV-1 is usually a discarded side-stream from the blood plasma fractionation of human serum albumin. It holds 90% of PC in plasma, but is very cheap. Preliminary study of PC purification from the Cohn Fraction IV-1 showed 16% purification yield using mini-Mab immobilized NHS-activated Sepharose. The economic analysis for PC purification using mini-Mab showed that the overall process was found to be tens of times cheaper than that using Mab.

1. INTRODUCTION

Deficiency of protein C (PC) in blood plasma may cause dangerous blood clot formation in various parts of the body, hindering oxygen and nutrient transport to tissue. PC may be a natural therapeutic for patients with PC deficiency, and also for various thromboembolisms and sepsis patients, with few side effects. Currently, monoclonal antibody (Mab) is used to purify PC from blood plasma. To reduce PC purification costs,

* Samin Rezania, Doh G. Ahn, and Kyung A. Kang, Department of Chemical Engineering, University of Louisville, Louisville, KY 40292.

E. coli cell line that can produce a variable region of an antibody (mini-antibody; mini-Mab) against PC was generated as an alternative ligand to Mab[1].

Currently, the mini-Mab is purified using protein A chromatography. This process is rather costly and the yield is only 33%[2]. Metal affinity chromatography (IMAC)[3, 4, 5] was previously investigated as an economic alternative to protein A. In our previous study by Rezania et al.[6], using the purified PC specific mini-Mab and Cu^{+2}-IDA Sepharose, a yield of 55% at pH 6.4 and 40% at pH 7.4 was obtained. As the next step the mini-Mab purification from *E. coli* cell culture broth was investigated. The effect of media components on mini-Mab adsorption to Cu-IDA was studied.

Cohn et al.[7] have developed a method to fractionate blood plasma proteins, on a large scale. Cohn Fraction IV-1 is a byproduct of Cohn Fractionation for human serum albumin purification and usually considered to be a waste; 90% of PC in plasma is included in this fraction[7]. An inexpensive PC source, Cohn Fraction IV-1, was used to obtain PC purified by the mini-Mab immobilized affinity matrix.

2. MATERIALS AND METHODS

2.1. Mini-Antibody purification using IMAC

Materials and methods for the mini-Mab production and its purification using Protein A Sepharose™ (Sigma; St. Louis, MO) are described in the paper by Korah et al[2]. Purification of the mini-Mab using Cu-IDA from *E. coli* cell culture broth was performed following the method described by Rezania et al[6]. Briefly, metal ion was loaded to the matrix using 0.1 M copper sulfate (Sigma) in water. 25 ml of cell culture broth was centrifuged at 4000 rpm for 20 minutes and the supernatant was loaded to the column containing 1 ml Cu-IDA Sepharose (binding capacity, 12 mg protein/ml matrix). The non-adsorbed mini-Mab was washed with the equilibrium/washing (E/W) buffer (0.02 M phosphate buffer). Five column volumes (CV) of 0.5 M imidazole in 0.02 M phosphate buffer was used to elute the bound mini-Mab. NaCl at 0.5 M was included in all process buffers to minimize non-specific binding. The mini-Mab quantification was performed by the ELISA, using anti-c-myc antibody[2], 9E10 (Santa Cruz Biotechnology; Santa Cruz, CA). 96 well microtiter plates (Nalgen Nunc International; Roskilde, Denmark) were used for ELISA.

2.2. PC purification

NHS-activated Sepharose™ 4 Fast Flow (NHS Sepharose™) and diethylaminoethyl (DEAE) Sepharose™ Fast Flow (DEAE Sepharose™) were purchased from Amersham Biosciences (Piscataway, NJ). PC and Cohn Fraction IV-1 paste were provided by the American Red Cross (Rockville, MD). 2-amino-2-(hydroxymethyl)-1,3-propanediol (Tris base), $NaHCO_3$, Na_2CO_3, phosphate buffered saline (PBS), Tween 20, horseradish peroxidase-conjugated goat anti mouse IgG, o-phenylenediamine dichloride (OPD) solution containing urea hydrogen peroxide, human serum albumin, and bovine serum albumin (BSA) were purchased from Sigma-Aldrich. The materials for human blood factors and the antibodies were used as described by Rezania et al[6].

Cohn Fraction IV-1 paste was dissolved in sodium citrate buffer at pH 6.0, to reduce the activity of protease present in plasma, with a dilution ratio of 1 to 20. Then it was centrifuged at 4000 rpm for 20 minutes and was filtered with filters at three different pore sizes (0.8, 0.45, and 0.2 μm), sequentially. Using the mini-Mab immobilized NHS Sepharose™; the PC purification from Cohn Fraction IV-1 was performed by the method described by Kang et al[8]. The amounts of the mini-Mab and PC were quantified by ELISA.

3. RESULTS AND DISCUSSION

3.1. Mini-Antibody purification using IMAC

In our previous study of the pH effect on mini-Mab purification with Cu^{+2}-IDA using pure mini-Mab, at the adsorption pH 6.4, a purification yield of 55% was achieved and at pH 7.4, a 40% yield with extensive metal leaching was observed[6].

As a next step, the pH effect on the adsorption of the mini-Mab in the *E. coli* reactor broth to Cu-IDA was investigated. The average mini-Mab concentration in cell culture broth is 0.5 mg/ml. Twenty five (25) ml reactor broth without cells, containing 12 mg mini-Mab was applied to 1 ml Cu^{+2}-IDA gel matrix. Two pHs (6.4 and 7.4), which gave us good results with the pure mini-Mab, were tested for this study. As shown in Figure 1, at pH 6.4, 91% of mini-Mab was washed away and nothing eluted from this matrix with this performing condition. At pH 7.4, 87% of mini-Mab was washed away and only 1% of mini-Mab purification yield was obtained.

Mini-Mab production media (2XTY medium) has 16 grams of bactotryptone, 10 grams of yeast extract and 5 grams of sodium chloride (NaCl) per 1 liter. From the results obtained, it was speculated that other molecules in the media may have higher affinity for the metal ion than the mini-Mab. Yeast extract and bactotryptone were the major components in the media, and bactotryptone has 13% of nitrogen and may compete with histidine for copper ions on the matrix. Therefore, the effect of media components on mini-Mab adsorption to copper ion was studied. Bactotryptone and yeast extract were examined for their effects on mini-Mab purification with IMAC (Figure 2).

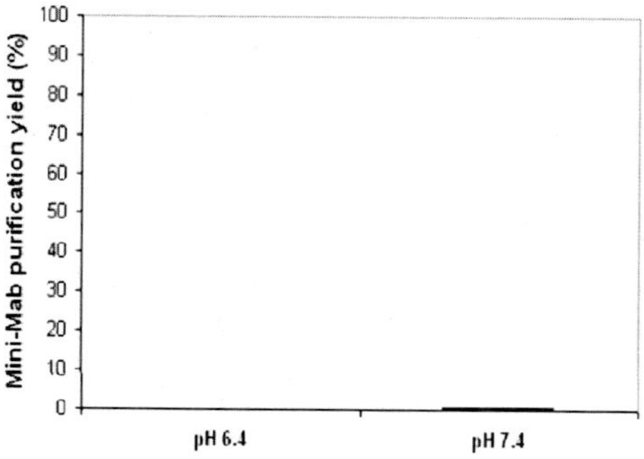

Figure 1. Mini-Mab purification from *E. coli* cell culture broth at pH 6.4 and 7.4.

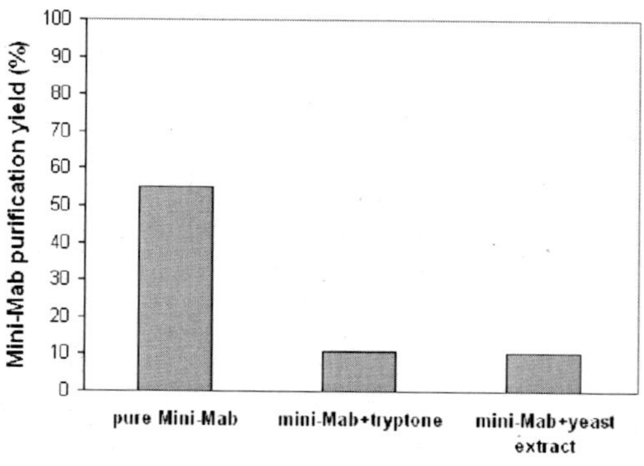

Figure 2. Effect of media components on mini-Mab adsorption to Cu-IDA.

0.5 mg of pure mini-Mab in 1 ml of the bactotryptone solution at pH 6.4 was applied to Cu-IDA matrix. ELISA results showed an 11% purification yield. For yeast extract, the same experiment was repeated using 0.5 mg pure mini-Mab in 1 ml of yeast extract solution at pH 6.4 and a 10.5% purification yield was obtained. These results showed that some components of media (yeast extract and bactotryptone) had a significant influence on mini-Mab adsorption to Cu-IDA.

Currently, more studies are investigating pretreatment of the reactor broth before applying to the IMAC matrix to reduce the effect by media components. Selection of other metal-chelate combinations that have higher affinity to the mini-Mab and maximize the mini-Mab adsorption is another ongoing study.

3.2. PC Purification from Cohn Fraction IV-1

PC purification from the Cohn Fraction IV-1 was investigated using the purified mini-Mab by protein A. It has been reported that 1 kg of Cohn Fraction IV-1 paste contains approximately 100 mg of PC and over 90% of PC in plasma[1]; however, PC homologues in Cohn Fraction IV-1 have not been quantified. The amount of PC and PC homologues in Cohn Fraction IV-1 were quantified by ELISA. The amount of PC was determined to be 100 mg/kg paste; the amount of factors VII and IX was much less than that of PC. The amount of factors II and X was approximately 10 and 2.5 times, respectively, of that of PC. The ratio between PC and factor II was 1 to 12, and PC and human serum albumin (HSA) was 1 to 240. Using 1 ml of NHS matrix, PC purification from Cohn Fraction IV-1 was performed using immuno-affinity chromatography. 4 ml of filtrate containing 15 μg of PC was applied to matrix. pH of the washing/adsorption buffer and Cohn Fraction IV-1 was adjusted to 6. PC and PC homologues in washing and elution steps were quantified using ELISA (Table 1). The PC purification yield was 16%. For the other homologous proteins, less than 0.4% was included in the PC eluate. This PC yield of 16% from the Cohn Fraction IV-1 is also almost a half of the yield from PC purification from pure PC and the PC homologues mixture at their physiological ratio (30%); data not shown. In the eluate, the actual amount of factor II was found to be almost half of the actual PC amount although it was just 0.3% of the initial amount. Since the main difference between this source and the mixture of PC and homologues is a large amount of HSA, the effect of HSA on mini-Mab adsorption was further studied.

3.3. Effect of HSA on PC Purification Performance

There was a reduction in the purification yield for the sample of the mixture of PC and PC homologues compared with the sample with PC only. In the presence of a large quantity of other molecules, PC molecules do not get transported to the mini-Mab well. Although HSA in Cohn Fraction IV-1 is only 1/30 of that in plasma, the amount is still 240 times greater than that of PC. Therefore, it was speculated that this large amount of HSA may hinder the diffusion of PC molecules.

Table 1. Amount of PC and PC homologues in purification process from Cohn Fraction IV-1.

		Amount of factors				
		PC	FII	FVII	FIX	FX
Yield	μg (%)	2.4±0.3 (16±1)	1.2±0.1 (0.3±0.1)	0 (0)	0.003±0.001 (0.3±0.1)	0.01± 0.003 (0.4±0.1)

Table 2. Effect of HSA on PC purification performance using the mixture of PC, PC homologues, and HSA.

			PC	FII	FVII	FIX	FX	HSA
Elution	W/O HSA	%	30.4	1.9	3.6	1.2	0.6	-
		μg	(6.1)	(8.4)	(0.1)	(0.2)	(0.2)	
	W/ HSA	%	12	0.1	17.2	0.2	0.6	0.02
		μg	(2.4)	(0.6)	(0.5)	(0.02)	(0.2)	(41)

The effect of HSA in the source material on PC purification was systematically studied. Each factor in washing and elution steps was quantified by ELISA (Table 2).

The PC purification yields were shown to be 12% and 25% for the samples with and without HSA, respectively, showing that, with HSA in the source, the yield is significantly reduced. Therefore, ion exchange chromatography (DEAE) was performed to remove HSA from Cohn Fraction IV-1 and over 99% of HSA was removed.

3.4. PC Purification from the Eluate of DEAE Chromatography of Cohn Fraction IV-1

PC purification from the eluate of DEAE was applied to the mini-Mab immobilized NHS matrix. Approximately 20 μg of PC and 4 μg of HSA were in the source. PC, PC homologues, and HSA in elution step were quantified using ELISA (Table 3).

The result showed that approximately 25% of the PC purification yield was obtained, which is a 9% higher yield compared with the yield (16%) using Cohn Fraction IV-1. The overall PC purification yield from Cohn Fraction IV-1 was 16% combining the yield using DEAE anion exchange chromatography from Cohn Fraction IV-1 (65%) and the yield by mini-Mab immobilized NHS matrix using the eluate of DEAE chromatography (25%). It was similar to the PC purification yield from Cohn Fraction IV-1 directly, although the FII amount is half of the product amount with the DEAE process.

Table 3. PC purification from the DEAE eluate of Cohn Fraction IV-1 using mini-Mab immobilized NHS matrix.

		PC	FII	FVII	FIX	FX	HSA
Amount in the source	(μg)	20	173±7	0.2±0.1	1.5±0.1	24±4	4±1.4
Amount in the eluate	μg (%)	4.9±0.1 (25±0.3)	2.5±1.4 (1.5±0.9)	0.1±0.1 (18±25)	0 (0)	0.1±0.02 (0.2±0.1)	0.3±0.01 (10±4)

Table 4. Economic analysis of PC purification using mini-Mab and regular Mab.

	Mini-Mab	Mab
Production cost	1	1000
Antibody purification	2	1
PC purification	4.5	1
Overall ratio	9	1000

A brief economic analysis for PC purification using mini-Mab showed that the overall process was found to be approximately 100 times more cost effective than that using Mab (Table 4).

4. CONCLUSIONS

PC is a natural anticoagulant, antithrombotic, and anti-inflammatory agent that can be used for the treatment of various thromboembolic complications. To reduce the cost of PC purification, Cohn Fraction IV-1 (which is an inexpensive source for PC) and using the cost-effective mini-Mab that can be produced by *E. coli*, was investigated for the PC purification process. IMAC was tested for the mini-Mab purification, as an inexpensive alternative to protein A. For mini-Mab purification from *E. coli* cell culture broth, over 80% of mini-Mab was washed away during the adsorption stage. Bactotryptone and yeast extract in media were found to have a significant effect on mini-Mab adsorption to Cu^{+2}-IDA.

When Cohn Fraction IV-1 was used, the PC purification yield using the mini-Mab immobilized NHS Sepharose^{TM} was 16%, lower than the case with pure PC (65%). A large amount of HSA in the source was found to influence the PC purification performance. DEAE chromatography was able to remove 99% of HSA, and 65% of PC was retained in the eluate. Using DEAE eluate as source material for PC purification, a 25% PC purification yield was achieved, much higher than that without the DEAE process (16%). The overall yield for PC purification from Cohn Fraction IV-1 using DEAE chromatography and mini-Mab affinity chromatography was also found to be 16%.

5. ACKNOWLEDGEMENTS

The authors thank the American Red Cross for supplying protein C, and Dr. Michael Sierks at the Arizona State University for providing *E. coli* colonies producing PC mini-Mab.

6. REFERENCES

1. H. Wu, G. N. Goud, and M. R. Sierks, Artificial Antibodies for Affinity Chromatography of Homologous Proteins: application to blood proteins, *Biotechnol. Progr.* **14**, 496-499 (1998).
2. L. K. Korah and K. A. Kang, Prelimanary Study for the Protein C Purification Using Mini-Antibodies Produced from Recombinant *E. coli*, *Oxygen Transport to Tissue XXV.* Thorniley, Harrison, and James (Kluwer Academic/Plenum Publishers, 2003), pp. 171-176.
3. L. Shukla, J. Boldman, S. Waugh, Process Characterization for Metal Affinity Chromatography of an Fc Fusion Protein, *Biotechnol. Appl. Biochem.* **34,** 71-80 (2001).
4. P. Verheesen, C. T. Verrips, and J. W. Haard, Beneficial Properties of Single-domain Antibody Fragments for Application in Immuno affinity Purification and Immuno-perfusion Chromatography, *Biochimicaet Biophysica. Acta.* **1624**, 21-28 (2003).
5. J. L. Casey, P. A. Keep, K. A. Chester, R. H. J. Begent, Purification of Bacterially Expressed Single Chain Fv Antibodies for Clinical Applications Using Metal Chelate Chromatography, *J. Immunol. Meth.* **179,** 105-116 (1995).
6. S. Rezania, D. G. Ahn, K. A. Kang, Cost Effective Metal Affinity Chromatography for Protein C Specific, Mini-Antibody Purification, Proceedings of the Annual 2004 ISOTT meeting [In press].
7. A.N.H. Creager, 'What Blood Told Dr Cohn': World War II, Plasma Fractionation, and the Growth of Human Blood Research, Stud. Hist. Phil. Biol. Biomed. Sci. **30**, 377-405.
8. K. A. Kang, D. Ryu, W. M. Drohan and C. L. Orthner, Effect of Matrices on Affinity purification of Protein C. *Biotechnol Bioeng.* **39**, 1086-1096 (1992).

THE ROLE OF ATP SENSITIVE CHANNELS IN INSULIN SECRETION AND THE IMPLICATIONS IN PERSISTENT HYPERINSULINEMIC HYPOGLYCAEMIA OF INFANCY (PHHI)

J.H. Shah*#, D.J. Maguire*, D. Brown#, A.Cotterill# [1]

Abstract: Persistent Hyperinsulinemic Hypoglycaemia of Infancy (PHHI) is a metabolic syndrome of unregulated insulin secretion. It is a heterogenous disease with causes linked to mutations of the ATP sensitive potassium channels of the β cell, as well as to metabolism in the β cell. 5 candidate genes – ABCC8, KCNJ11, GCK, GLUD1 and SCHAD have been implicated in the disease so far, however the aetiology of the disease remains unknown in up to 50% of all patients. We genotyped 43 subjects with PHHI (20 surgically treated and 23 medically treated) for disease associated mutations in the candidate genes. Mutations on ABCC8 were identified in 16 of the 20 (80%) of the surgically treated patients. One putative mutation was identified in the medically treated cohort. The polymorphism E23K on KCNJ11 that is associated with NIDDM was differentially distributed in the 2 cohorts. We discuss the mutations identified, emphasise the importance of the K-ATP channel in physiological processes and discuss the possibility that the disease is caused by mutations in other genes associated with insulin release, glucose metabolism in the β cell or β cell apoptosis and survival. We propose that these processes must be explored in order to further our understanding of PHHI.

1. INTRODUCTION

In recent years it has become clear that ATP-Sensitive Potassium Channels (K-ATP channels) have a key role in the physiology of cells. They play a vital role in linking cell metabolism and ATP production to cellular electrical activity[1, 2]. They are involved in

* School of Biomolecular and Biomedical Sciences, Griffith University, Nathan, Aus
# Mater Children's Hospital, Dept of Paediatric Endocrinology, Brisbane, Aus

glucose sensing in the pancreatic beta cells and in the brain, in regulation of vascular smooth muscle tone, in ischaemic preconditioning of cardiac myocytes and in seizure protection[1]. Defects of the channel itself or in its regulation can cause human disease[3].

K-ATP channels were first identified in cardiac myocytes and in pancreatic beta cells[4,5]. In the pancreatic beta cells the K-ATP channels are part of an ionic mechanism that couples glucose metabolism to insulin release. These β cell K-ATP channels respond to intracellular nucleotides, especially ATP and Mg^{2+}ADP in an antagonistic manner; ATP acts as a channel blocker, while Mg^{2+}ADP acts as a channel opener[3]. The difference in response to the phosphorylated nucleotides has been attributed to the fact that the β cell expresses a K-ATP channel complex formed by subunits belonging to at least two distinct families of proteins [3,6]. The potassium selective pore is formed by the inwardly rectifying potassium ion channel Kir6.2 and the ATP binding cassette super family receptor by SUR1[6]. The subunits are encoded by the genes KCNJ11 and ABCC8, respectively, that lie adjacent to each other at 11p15.1 [4, 7, 8, 9].

K-ATP channels are arranged in a 4:4 octameric complex of 4 Kir6.2 subunits arranged around a central pore coupled with 4 SUR1 subunits[9]. Each subunit carries an endoplasmic reticulum retention signal – an RKR motif, which ensures that only correctly assembled channels are transported to the cell surface [10].

Ion channelopathies have been described in many cell types including neurons, myocytes and endocrine cells. Hyperinsulinism of Infancy (HI) is a rare, potentially lethal condition of the new born and early childhood[6,11]. The specific metabolic syndrome Persistent Hyperinsulinemic Hypoglycaemia of Infancy (PHHI), is a metabolic syndrome characterized by the hypersecretion of insulin in the presence of often severe hypoglycaemia. Patients usually present in the first 72 hours of life with symptoms of hypoglycaemia that include feeding problems, loss of consciousness and seizures. PHHI is biochemically associated with elevated insulin, low glucose and usually absence of ketosis. In severe cases coma or death may occur[6,11]. Up to 50% of patients with PHHI suffer from long-term effects of neurological impairment[11,12]. Treatment options usually include the use of potassium channel openers such as diazoxide and octreotide, however in some cases, these may not be effective, and in such cases a partial or subtotal pancreatectomy may be necessary [12].

PHHI is a clinically and genetically heterogeneous disease which often renders it difficult to diagnose and treat. The various causes of PHHI share a common target protein - the K-ATP channel[6]. To date, mutations in 5 genes have been associated with PHHI – ABCC8 and KCNJ11 that encode the K-ATP channel subunits, glucokinase (GCK) that encodes the enzyme that catalyses the rate limiting enzyme in the metabolism of glucose in the β cell [13], glutamate dehydrogenase (GLUD1) that encodes the enzyme that catalyses the oxidative deamination of glutamate leading to the production of α-keto glutarate which is an essential substrate in the krebs cycle[14] and the short chain fatty acid hyroxyacyl dehydrogenase (SCHAD) that plays a vital role in the beta oxidation of short chain fatty acids in the mitochondria[15,16].

Normally, as glucose enters the β cell it is metabolized to pyruvate via the glycolytic pathway; subsequently it enters the Krebs cycle which results in the production of ATP. The ATP produced results in an increase of the intracellular ATP:ADP ratio. The ATP binds to the KIR6.2 subunit of the channel causing it to close, thus depolarizing the cell membrane[17]. The depolarization triggers the voltage dependent calcium channel to open initiating an influx of calcium,- thus beginning the cascade that stimulates in the secretion of insulin from the β cell. As ATP production decreases and the ATP:ADP ratio decreases, the Mg^{2+}ADP in the cell binds to the SUR1 subunit of the channel triggering it to open, and the cell regains its electric potential – thus stopping the release of insulin. Mutations associated with HI within the K-ATP channel subunits result in the channel either not being expressed, or being non functional – thus preserving the β cell in a depolarized state (due to actions of the Na-K ATPase)– this leads to the constitutive release of insulin resulting in hypoglycaemia [6].

The aim of our study has been to understand the molecular aetiology of the disease and the role of the K-ATP channel in an Australian cohort.

2. METHODOLOGY

A cohort of 43 patients diagnosed with PHHI has been assembled over the last 18 years. Patients within this cohort have been segregated by their response to medical treatment with Diazoxide. Those that responded to diazoxide were assumed to have a milder form of the disease and were classified as the "medical" cohort. Those that required surgery for management were assumed to have a more severe form of the disease – this was the "Surgical' Cohort.

The patients were genotyped for selected genes in an attempt to identify the disease causing mutation. GCK, GLUD1, KCNJ11 and ABCC8 were analysed in the medical cohort.. KCNJ11 and ABCC8 were sequenced in the surgical cohort (it was assumed that the lack of response to medical treatment implied a severe disease and a channel defect).

Genotyping was performed using genomic DNA extracted from leukocytes. The coding regions of the selected genes were amplified by the Polymerase Chain Reaction (PCR) and then sequenced using the "Big Dye Terminator Cycle Sequencing Ready" reaction. Primers were designed to target the exon as well as splice sites. All coding regions and splice sites of GCK, ABCC8 and KCNJ11 were sequenced, exons 6, 7, 8, 11 & 12 associated with allosteric regulation were sequenced in GLUD1 [14].

Sequences were analysed using the program Chromas and compared against consensus sequences available from NCBI (Accession Numbers:ABCC8-51470970, KCNJ11-51470970, GCK-15967157, GLUD1-4885280.)

3. RESULTS

No pathogenic mutations were identified in the GCK, GLUD1 and KCNJ11 genes. Several mutations were identified in ABCC8 in the surgical cohort – some of which have been previously reported to be associated with PHHI, and some of which are novel and believed to be associated with the disease.

Twenty surgical patients were genotyped for ABCC8, mutations were identified in 16 of them (80%). Of the 16 patients, 4 had compound heterozygous mutations while 12 had mutations inherited from either parent. Of the 20 mutations identified (heterozygous and homozygous), 9 have been previously reported and 11 are novel putative mutations. The mutations are spread throughout the coding region of the gene, but some are situated at splice sites in the intronic regions. The two nucleotide binding folds of the channel that bind to ATP are hot spots for mutations associated with PHHI. Only one putative mutation was identified in the 43 patients of the medical cohort.

The analysis of KCN11 showed the uneven distribution of polymorphisms that are associated with Non Insulin Dependent diabetes[18]. The polymorphisms E23K and I337V were seen in 64% of medical patients, but only in 37% of surgical patients and normal subjects. The A190A variant that is known to add to the maximal load of variance in diabetes patients[18] was seen in approximately 30% of the surgical and normal subjects and in 74% of the medical patients.

4. DISCUSSION

Genetic analysis of the 4 genes widely believed to be involved in the development of PHHI have shown pathogenic mutations in the ABCC8 gene in the two Australian cohort. Genotyping KCNJ11 in both cohorts showed sequence variants that had significantly different allele frequencies in the 2 cohorts ($p<0.0001$). Analysis of GCK and the allosteric domain of GLUD1 showed naturally occurring gene variants. SCHAD has not been targeted as mutations in the gene have so far only been reported in a Norwegian cohort [15]. There are currently more than 60 mutations reported in ABCC8- the mutations are distributed across the 39 exons of the gene, however the exons encoding the 2 nuclear binding folds (NBF's) are hot spots for mutations[19]. The NBF's have ATPase activity and are involved in the sensing of intracellular ATP concentration[20]. Mutations in this region would affect the subunits ability to bind to ATP, and result in a non-functional channel that could lead to hypoglycaemia. Four putative mutations identified are likely to result in the production of a truncated protein that might result in loss of K-ATP channel function.

Analysis of the sequence variants seen in KCNJ11 might have implications for the patients' susceptibility to diabetes, as well as their response to medical treatment. Hansen et al[18] has shown that carriers of the E23K, I337V and A190A load of variants have increased susceptibility to diabetes. While Hani et al. (1998) and others have shown that carriers of the K/K variant (at codon 23) have impaired insulin secretion[21,22]. The gene variants E/K or K/K are believed to alter diabetes susceptibility by increasing the threshold of intracellular ATP necessary for insulin secretion[18]. The difference in the distribution between the surgical and medical patients implies that the variants may play a role in the channel folding and thus affect the carriers' ability to respond to drugs used in the treatment of PHHI.

Despite extensive genetic analysis of patients diagnosed with PHHI, the molecular aetiology remains unknown in 40-50% of these cohort and others[16,23]. There are several possible explanations for this, the most obvious being that the disease causing mutation is located on some other gene involved in insulin release or in the glucose sensing pathway. A mutation located in a region of the gene that we did not target such as the promoter region shown to be defective in two other studies might also account for this finding[24]. Hyperinsulinism may also be a by-product of some other underlying metabolic condition such as a congenital disorder of Glycosylation (CDG) or a mitochondrial myopathy.

Initial studies by our group, and others[25] have shown an increased rate of proliferation and apoptosis in patients with PHHI as compared to a normal control cohort. In cases of PHHI associated with a K-ATP channel function defect, there is a permanent state of depolarization of the cell membrane, which causes the voltage dependent calcium channels to remain open[17]. The resulting influx of calcium is responsible for the constitutive secretion of insulin; however it is also believed to play a role in initiating apoptosis via the caspase pathway[26]. The mechanisms involved in the initiation of apoptosis in HI are not fully understood. Could the apoptosis also be a result from the cell suffering from a hypoxic trauma? While the influx of calcium accounts for the increased apoptosis, it does not explain why it is seen only at certain times – the entire pancreas is not affected. Is the cell in crisis as it is suffering from fuel deprivation combined with over exertion, resulting in cell hypoxia and increased rate of apoptosis? Figure 1 summarizes some of the mechanisms that might be associated with the increased rate of apoptosis seen in Hyperinsulinemic beta cells.

In continuing efforts to understand the aetiology of PHHI in these Australian cohorts, work will continue to elucidate the genetic abnormality related to the disorder and understand its macromolecular implications. An important priority is to establish a genotype – phenotype correlation in PHHI to better diagnose and treat patients.

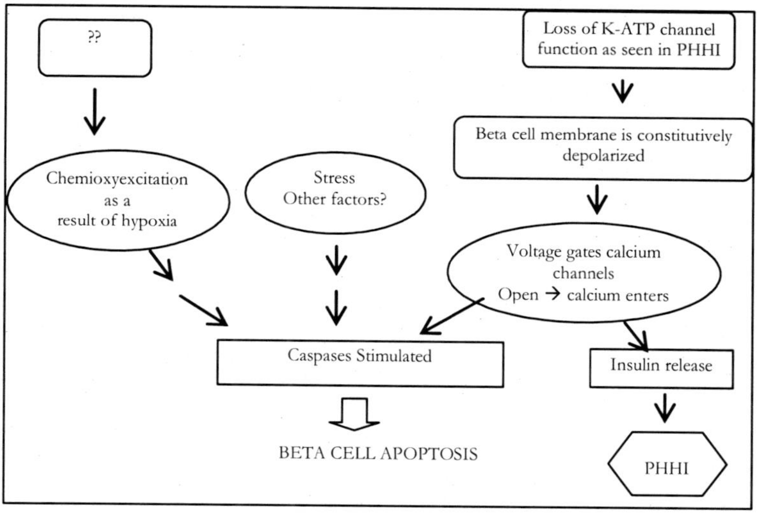

Figure 1. A flow diagram showing the possible connection between PHHI and increased beta cell apoptosis.

5. ACKNOWLEDGEMENTS

The authors would like to acknowledge members of the Mater Children's Hospital PHHI research group, Dr. David Cowley, Dr. Ross Walker, Dr. Frank Bowling, Dr. Ristan Greer, Dr Michele Jack and Daphne Liaskou.

6. REFERENCES

1. Dabrowski M, Tarasov A, Ashcroft FM. *Mapping the architecture of the ATP binding site of the K-ATP channel subunit KIR6.2.* The Journal of Physiology, 2004. **557**(2): p. 347-354.
2. Ashcroft FM. *Adenosine 5'-triphosphate-sensitive potassium channels.* Annu Rev Neurosci, 1988. **11**: p. 97–118.
3. Seino S. and T Miki. *Physiological and pathophysiological roles of ATP sensitive potassium channels.* Progress in biophysics and molecular biology,2003. **76**: p. 133-176.
4. Noma A. *ATP-regulated K^+ channels in cardiac muscle.* Nature, 1983. **305**(5930): p. 147-8.
5. Cook DL, and Hales CN.*Intracellular ATP directly blocks K^+ channels in pancreatic B-cells.* Nature, 1984. **311**(5983): p. 271-3.

6. Dunne MJ, Cosgrove KE, Shepherd RM, Aynsley-Green A, Lindley KJ. *Hyperinsulinism in Infancy: from Basic Science to Clinical Disease.* Physiol Rev, 2004. **84(**1): p. 239-275.
7. Aguilar-Bryan L, Clement JP 4th, Gonzalez G, Kunjilwar K, Babenko A, Bryan J. *Toward understanding the assembly and structure of KATP channels.* Physiol Rev, 1998. **78**(1): p. 227–245.
8. Inagaki N, Gonoi T, Clement JP 4th, Namba N, Inazawa J, Gonzalez G, Aguilar-Bryan L, Seino S, Bryan J. *Reconstitution of IKATP an inward rectifier subunit plus the sulfonylurea receptor.* Science, 1995. **270**(5239): p. 1166.
9. Shyng SL and Nichols CG. *Octameric stoichiometry of the KATP channel complex.* Journal of General Physiology, 1997. **110**(6): p. 655–664.
10. Zerangue N, Schwappach B, Jan YN, Jan LY. *A new ER trafficking signal regulates the subunit stoichiometry of plasma membrane KATP channels.* Neuron, 1999. **22**(3): p. 537–548.
11. Aynsley-Green A, Hussain K, Hall J, Saudubray JM, Nihoul-Fekete C, De Lonlay-Debeney P, Brunelle F, Otonkoski T, Thornton P, Lindley KJ. *Practical management of hyperinsulinism in infancy.* Arch Dis Child Fetal Neonatal Ed. Arch Dis Child, 2000. **82**(2): p. F98–F107.
12. Jack MM, Greer RM, Thomsett MJ, Walker RM, Bell JR, Choong C, Cowley DM, Herington AC, Cotterill AM. *The outcome in Australian children with hyperinsulinism of infancy: early extensive surgery in severe cases lowers risk of diabetes.* Clin Endocrinol (Oxf), 2003. **58**(3): p.355-64.
13. Davis EA, Cuesta-Munoz A, Raoul M, Buettger C, Sweet I, Moates M, Magnuson MA, Matschinsky FM. *Mutants of glucokinase cause hypoglycaemia- and hyperglycaemia syndromes and their analysis illuminates fundamental quantitative concepts of glucose homeostasis.* Diabetologia, 1999. **42**(10): p. 1175-1186.
14. Stanley CA, Lieu YK, Hsu BY, Burlina AB, Greenberg CR, Hopwood NJ, Perlman K, Rich BH, Zammarchi E, Poncz M. *Hyperinsulinism and Hyperammonemia in Infants with Regulatory Mutations of the Glutamate Dehydrogenase Gene.* N Engl J Med, 1998. **338**(19): p. 1352-1357.
15. Molven A, Matre GE, Duran M, Wanders RJ, Rishaug U, Njolstad PR, Jellum E, Sovik O. *Familial Hyperinsulinemic Hypoglycemia caused by a defect in the SCHAD enzyme of mitochondrial fatty acid oxidation.* Diabetes, 2004. **53**(1): p. 221-227.
16. Molven A, Rishaug U, Matre GE, Njolstad PR, Sovik O. *Hunting for a Hypoglycemia Gene: Severe Neonatal Hypoglycemia in a Consanguineous Family.* Am J Med Genet, 2002. **113(**1): p. 40-46.
17. Ghosh A, Ronner P, Cheong E, Khalid P, Matschinsky FM. *The role of ATP and free ADP in metabolic coupling during fuel-stimulated insulin release from islet {beta}-cells in the isolated perfused rat pancreas.* J Biol Chem, 1991. **266**(34): p. 22887–22892.
18. Hansen L, Hansen T, Urhammer SA, Clausen JO, Pedersen O. *Amino acid polymorphisms in the ATP-regulatable inward rectifier Kir6.2 and their relationships to glucose- and tolbutamide-induced insulin secretion, the insulin sensitivity index, and NIDDM.* Diabetes 1997. **46**: p.508-512,
19. MutDB, Indiana School of Medicine. (15th Aug 2005); http://mutdb.org.
20. Uhde I, Toman A, Gross I, Schwanstecher C, Schwanstecher M. *Identification of the potassium channel opener site on Sulfonylurea Receptors.* J Biol Chem, 1999. **274**(40): p. 28079-28082.
21. Hani EH, Boutin P, Durand E, Inoue H, Permutt MA, Velho G, Froguel P. *Missense mutations in the pancreatic islet beta cell inwardly rectifying K+ channel gene (KIR6.2/BIR): a meta-analysis suggests a role in the polygenic basis of Type II diabetes mellitus in Caucasians.* Diabetologia, 1998. **41**(12): p. 1511-1515.
22. Elbein SC, Sun J, Scroggin E, Teng K, Hasstedt SJ. *Role of Common Sequence Variants in Insulin Secretion in Familial Type 2 Diabetic Kindreds: The sulfonylurea receptor, glucokinase, and hepatocyte nuclear factor 1{alpha} genes.* Diabetes Care, 2001. **24**(3): p. 472-478.
23. Kukuvitis A, Deal C, Arbour L, Polychronakos C. *An autosomal dominant form of familial persistent hyperinsulinemic hypoglycemia of infancy, not linked to the sulfonylurea receptor locus.* J Clin Endocrinol Metab, 1997. **82**(4): p. 1192-1194.
24. Gloyn AL, H.Y., Ashcroft SJ, Ashfield R, Gloyn AL, Hashim Y, Ashcroft SJ, Ashfield R,, *Association studies of variants in promoter and coding regions of beta-cell ATP-sensitive K-channel genes SUR1 and Kir6.2 with Type 2 diabetes mellitus (UKPDS 53).* Diabet Med, 2001. **18**(3): p. 206-212.
25. Kassem SA, Ariel I, Thornton PS, Scheimberg I, Glaser B. *Beta cell proliferation and apoptosis in the developing normal pancreas and in Hyperinsulinism of Infancy*. Diabetes, 2000. **49**(8): p. 1325-1333.
26 Voet, D. and J. Voet, *Biochemistry.* Second ed. 1995: John Wiley & Sons Inc.

TRIPTOLIDE ALTERS MITOCHONDRIAL FUNCTIONS

Ying Su, Shanmin Yang, Zhenyu Xiao, Wei Wang, Paul Okunieff, and Lurong Zhang*

Abstract: Triptolide (TPL), a small molecule purified from the herb *Tripterygium wilfordii*, has potential clinical application for suppression of chronic autoimmune disorders and inhibition of tumor growth. However, its mechanism of action is largely unknown. In this study, the effect of TPL on mitochondria was explored with a panel of molecular probes that detect the alteration of mitochondrial functions. When Lewis lung carcinoma (LLC) cells were treated with different doses of TPL for four hours, impaired mitochondrial functions were detected. This included an increased production of reactive oxygen species, the opening of the transition pore of mitochondria, the depolarization of the mitochondria membrane, the inhibition of the production of ATP and increased release of ATP as well as the induction of apoptosis. It is likely that by impairment of mitochondrial function, TPL exerts its inhibitory effect on growth of tumor and progression of inflammatory disease.

1. INTRODUCTION

Triptolide (TPL) was purified from the medicinal herb *Tripterygium wilfordii Hook F* in 1972. It is a small molecule (diterpenoid triepoxides) with a molecular weight of 360.4 Da.[1] TPL is being tested on a number of autoimmune and neoplastic diseases. TPL inhibits the growth of a variety of primary and secondary mouse tumors at very low dose (0.2-0.25 mg/kg/twice a week).[1-4] The clinical dose is also very low; 30 μg/kg/every other day is being used for leukemia patients with a preliminary response rate of 71% (personal communication). TPL has been used in a phase II clinical trial for arthritis in the USA with promising effect.[5-7] In addition, TPL blocks pulmonary arterial hypertension and neointimal inflammation in a rat model.[8] Finally, TPL reduces bleomycin-induced lung fibrosis.[9]

The mechanism of action of TPL is largely unknown, though several key molecular pathways are known to be altered in the presence of TPL in cell culture and animal models. For example, characteristic apoptotic features, including DNA fragmentation and chromatin condensation, are observed in tumor cells treated with TPL.[10] How apoptosis is triggered and which pathway predominates is unknown.

* Department of Radiation Oncology, University of Rochester Medical Center, Rochester, NY 14642 USA.
This work supported in part by (NIH/NIAID) AI067733-01.

TPL inhibits NFκβ transcription activity and results in up-regulation of IκBα mRNA expression.[11, 12] Other signaling molecules are also affected. The anti-inflammatory effect of TPL is associated with inhibition of the production of a panel of cytokines including interleukin-2 (IL-2), IL-6, IL-8, macrophage inflammatory protein-2α (MIF-2α), intercellular adhesion molecule-1, integrin β_6, vascular endothelial growth factor (VEGF), and granulocyte-macrophage colony-stimulating factor (GM-CSF).[12-14]

The mitochondria are involved in many aspects of apoptosis and inflammation; however, the effect of TPL on mitochondria has not been explored. In this study, we used a panel of probes to examine the mitochondrial alterations that occur upon treatment of Lewis lung carcinoma with different doses of TPL. The results suggest that the mitochondria may be one of the targets of TPL.

2. MATERIALS AND METHODS

2.1. Reagents and Cells

Lewis lung cancer cells (LLC) were cultured with 10% calf serum, 90% DMEM (Dulbecco's Modification of Eagle's Medium) at 37°C in an incubator with 5% CO_2. TPL with 99.9% purity was obtained from Eva Pharmaceuticals (Rochester, NY). CM-H2DCFDA {5-(and-6)-chloromethyl-2',7'-dichlorodihydro-fluorescein diacetate, acetyl ester}, JC-1, Calcein AM, Annexin V, propidium iodide (PI), and an ATP Determination Kit were purchased from Molecular Probes (Eugene, OR).

2.2. TPL Treatment of Cells

For flow cytometer analyses, 2 x10^5 LLC cells were cultured in 35 mm dishes to 80% confluence and then treated with TPL at different concentrations (0, 3, 10, 40 ng/ml). Four hours later, the cells were harvested and single cell suspensions of each treatment group were placed separately in 6 ml tubes with 1 ml culture media. The cells were allowed to rest for one hour to allow membrane to recover from damage that may have occurred during the harvesting process. The staining and flow cytometry were then performed as described below.

2.3. Analysis for Reactive Oxygen Species (ROS)

Single cell suspension (100 μl) from each treatment was incubated with CM-H2DCFDA (final concentration of 5 μM) at 37°C for 0.5 hour. The intensity of fluorescence was determined by flow cytometry at an excitation wave-length of 488 nm and emission wave-length of 520 nm. The unstained cells were used for autofluorescence. Higher levels of fluorescence indicated greater oxidation.

2.4. Analysis for Mitochondria Membrane Potential

Two μg of JC-1 in 30 μl of saline was added to 100 μl of single cell suspension treated with 0, 3, 10, or 40 ng/ml of TPL for 4 hours. The cells were further incubated at 37°C for 10 min, washed with PBS once, and then immediately subjected to flow cytometer analysis. Since membrane depolarization is indicated by a shift from high red to high green, the percentage of cells in the high or low red and green regions were measured and compared among the different treatment groups.

2.5. Analysis for Transition Pore of Mitochondria

Five μl of 2 μM Calcein AM was added to 100 μl of single cell suspension that had been treated for 4 hours with 0, 3, or 10 ng/ml TPL. Then 5 μl of 80 mM $CoCl_2$ was added to the tubes, which quenched the cytosolic fluorescence. If the transition pore is closed, fluorescence is maintained for a period of time in the mitochondria. The cells were incubated at 37°C for 15 min, washed once with PBS, and immediately analyzed with flow cytometry for the percentage of green fluorescent cells.

2.6. Analysis for Production and Release of ATP

LLC cells were cultured in 96 well black walled plates and treated with different concentrations of TPL with 30 μl of phenol red free media. Four hours later, 100 μl of reaction solution (containing D-luciferin and firefly luciferase) from the ATP Determination Kit (A-22066) was added to each well and incubated for 15 min. The intensity of bioluminescence that developed due to the reaction of ATP in the media was read with SpectraMax Gemini EM at an excitation wave-length of 488 nm and an emission wave-length of 560 nm. After the initial reading, 10 μl of 1% NP-40 in deionized H_2O was added to each well to release the intracellular ATP and the reading was repeated.

2.7. Analysis for Apoptosis

LLC cells treated with TPL for 4 hours at doses of 0, 3, 10 or 40 ng/ml were harvested and stained with Annexin V for 30 min and then with PI immediately before flow cytometry analysis. The percentage of cells that were Annexin V positive but PI negative was compared among the different treatment groups.

2.8. Statistical Analysis

The mean and standard error were calculated from the raw data and then subjected to Student's *t* test. The data for the vehicle alone (0 ng/ml of TPL) group and the groups treated with different doses of TPL were compared. A *P* value < .05 was regarded as statistically significant.

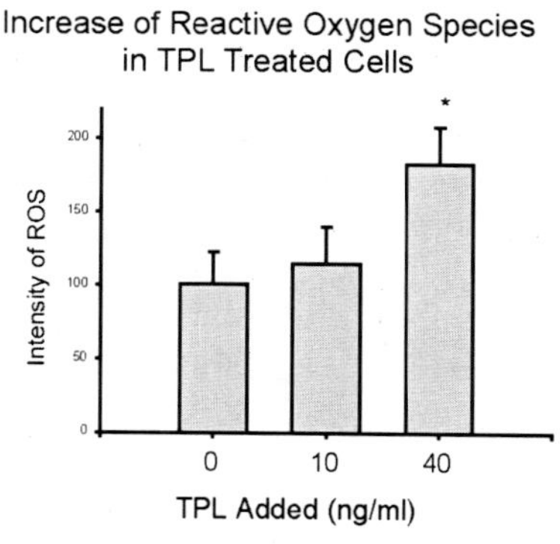

Figure 1. Increased ROS in TPL treated cells. The cells were treated with either media alone (as control) or 10 or 40 ng/ml of TPL for 4 hours. CM-H2DCFDA staining and flow cytometry were performed as described above. TPL at 40 ng/ml increased the production of ROS (*P< .05). 10,000 events were counted for this analysis. Similar results were obtained in separate experiments.

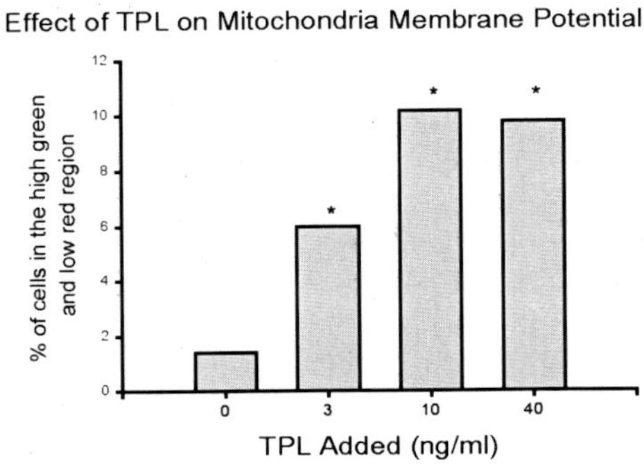

Figure 2. TPL depolarizes the mitochondria membrane. Cells were treated with different concentrations of TPL for 4 hour and then stained with JC-1 as described above. TPL increased the fraction of cells with a depolarized mitochondria membrane (*P< .01). 10,000 events were counted for this analysis. Repeat experiments produced similar results.

3. RESULTS

3.1. TPL Increases the Production of ROS

To examine the oxidation state of mitochondria after TPL exposure, CM-H2DCFDA was used. As this agent passes through plasma membrane, the acetate group is spontaneously removed by intracellular esterases and then oxidized with ROS produced by the TPL-stressed LCC cells. Oxidation is indicated by increasing fluorescence. The fluorescent intensity of LCC cells treated at 40 ng/ml TPL was much higher than that of untreated cells, indicating that the TPL-exposed cells had increased ROS (Figure 1).

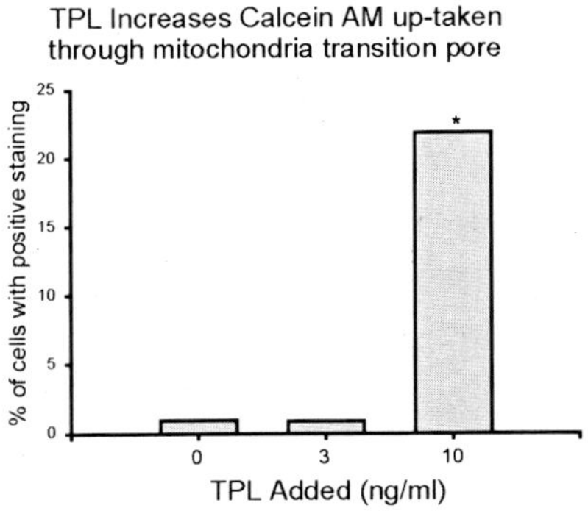

Figure 3. Effect of TPL on transition pore of mitochondria. Two μl of 2 μM Calcein AM was added to 100 μl LLC cells that were treated 4 hours without or with TPL, and then 5 μl of 80 mM of $CoCl_2$ was added to quench the cytosolic fluorescence. After incubation at 37°C for 15 min, the cells were washed with PBS and analyzed for the percentage of cells with mitochondrial fluorescence. 10,000 events were counted for this analysis. Similar result were obtained in separate experiments.

3.2. TPL Depolarizes the Mitochondria Membrane

We examined the mitochondrial membrane for depolarization using JC-1 staining followed by flow cytometry. Figure 2 shows that the percentage of cells in the low red and high green region was increased by TPL in a dose-dependent manner (P< .01). Maximum depolarization was obtained at 10 ng/ml TPL. The data suggest that TPL depolarizes the mitochondria membrane at a concentration that is clinically relevant.

3.3. TPL Opens the Transition Pore of Mitochondria

The depolarization of the mitochondrial membrane is an event that can result from the opening of the transition pore of the mitochondria. To see if this is the case, we used the Calcein AM assay. Acetylated Calcein AM can pass the membrane and then converts to a fluorescent dye after spontaneous esterase mediated hydrolysis. Cytosolic fluorescence is quenched by $CoCl_2$, but the mitochondrial fluorescence persists if the Ca^{2+} pore is closed. The percentage of cells with mitochondrial Calcein fluorescence was much higher (more than 10 fold) in cells exposed to 10 ng/ml TPL compared with the control or 3 ng/ml TPL-treated groups (Figure 3).

3.4. TPL Reduces Intracellular ATP and Increases Extracellular ATP

Since the mitochondria are the major producer of ATP, we speculated that the damage of mitochondria is likely to reduce the production of ATP. There was an increased release of ATP into the medium from TPL-treated cells (Figure 4A), while the intracellular level of ATP was greatly reduced (Figure 4B).

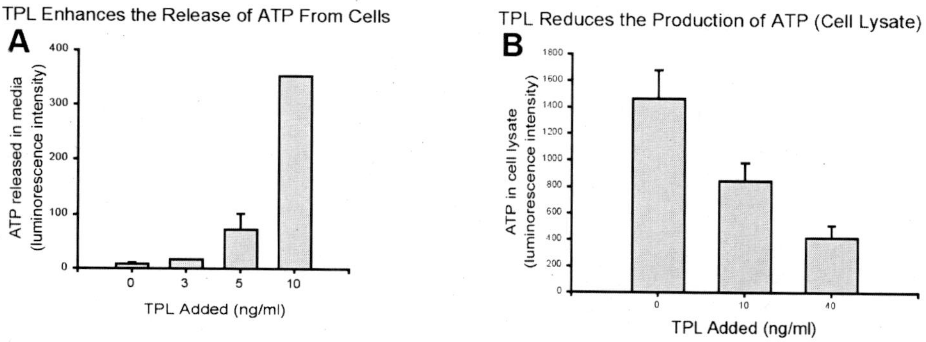

Figure 4. Effect of TPL on intracellular and extracellular ATP. LLC cells were cultured in 96 well plates with phenol red free media. Cells were exposed to TPL for 4 hours at a concentration of 0, 3, 5, 10, or 40 ng/ml. ATP in the media was first measured using a luciferase based ATP Determination Kit, and then 1% NP-40 was used to lyse the cell membrane. The total ATP was then measured using the same kit. The experiment was done in triplicate. The increased extracellular ATP and decreased intracellular ATP in LLC cells treated with 10 ng/ml of TPL was statistically significant (P< .01). This experiment was performed in triplicate.

3.5. TPL Triggers Apoptosis

Damage of mitochondria will trigger cell death by apoptosis or necrosis. Early mitochondria-related apoptosis can be detected by translocation of the phosphatidylserine (PS) from the inner to the outer leaflet of the plasma membrane.[15-17] The binding sites on PS become available for Annexin V. Annexin V is a 35-36 kDa Ca 2+-dependent phospholipid binding protein with a high affinity for PS. The LLC cells treated with TPL for 4 hours at the doses of 0, 3, 10 or 40 ng/ml were stained with Annexin V for 30 min and then with PI to distinguish the early apoptotic cells (Annexin V-positive and PI-negative) from necrotic cells (Annexin V and PI-positive). Apoptotic cells were unchanged from control at 10 ng/ml and were significantly increased at 40 ng/ml TPL.

4. DISCUSSION

While both studies in animal models of autoimmune and neoplastic disease, and preliminary clinical trials have demonstrated effectiveness of TPL, the underlying molecular mechanism of action of TPL is largely unknown. In other experiments (unpublished data), we have seen specific and prolonged accumulation of TPL in the mitochondria. In this study, we have therefore focused on effects of TPL on mitochondrial phenomenon. We show for the first time that TPL has the following effects on the mitochondrial function of LLC cells: **1)** it increased the production of reactive oxygen species at apoptosis-generating doses (Figure 1); **2)** it depolarized the mitochondria membrane in a dose-dependent manner reaching a maximum at a dose below that which caused substantial apoptosis (Figure 2); **3)** it opened the transition pore of the mitochondria at high dose consistent with the induction of apoptosis (Figure 3); **4)** it reduced the level of intracellular ATP and increased the level of extracellular ATP in a dose-dependent manner at doses lower than required for apoptosis (Figure 4); and **5)** it induced apoptosis at the highest TPL exposure levels (Figure 5).

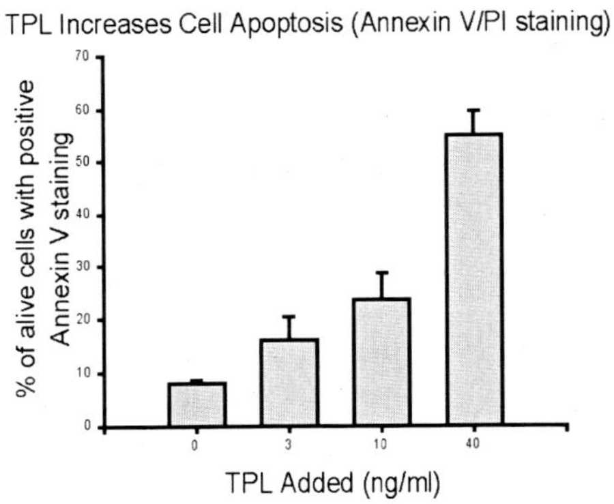

Figure 5. Effect of TPL on apoptosis. LLC cells were treated with TPL for 4 hours at doses of 0, 3, 10, or 40 ng/ml. Cells were then stained with Annexin V for 30 min and then with PI prior to the flow cytometer analysis. The apoptotic cells fraction increased along with TPL (P< .01). 10,000 events were counted for this analysis. Repeat experiments produced similar results.

These TPL effects on mitochondria related functions have therapeutic significance since they are achieved at a dose much lower than that used in the animal experimental models for anti-inflammation and anti-tumor effect.[1-9, 18-21] Circulating levels were also safely in the expected range based on the dose used in clinical trials.[5-7] Specifically, we found that TPL-related interruption of mitochondria functions occurs in a dose-dependent manner. Maximum effects are seen at concentrations of 10 to 40 ng/ml applied for 4 hours. Experimentally, when TPL is used, it is given i.v. at 200-300 μg/Kg. We have not yet performed detailed biodistribution studies, but simple calculations put this in vivo dose at approximately and order of magnitude above that used in cell experiments. We have found that TPL administrated at 4-6 μg/mouse/every other day can effectively inhibit the growth of all tumor xenografts tested, including breast cancer, prostate cancer, melanoma, gastric cancer and hepatoma.[2]

At these doses, TPL was well tolerated without obvious side-effects.[2] Clinically, at a dose of 30 μg/kg, TPL achieved a 71% response rate for leukemias (personal communication). At this dose, again assuming a water to weight conversion, the 10-40 ng/ml concentration is achievable in patients. Therefore, we believe that the TPL effects on mitochondrial functions can take place *in vivo* and might account for its biological effects.

While the initial target of TPL is unknown, the interruption of many mitochondrial functions likely causes the apoptosis and death of proliferating cells. For example, the opening of transition pores would allow Calcium ion inflow and precipitate apoptosis. Membrane depolarization would reduce the efficiency of ATP synthesis and thus precipitate necrosis. Leakage of ATP is expected from cells that have too little ATP to maintain the plasma membrane potential and integrity.[22] In addition, the pore opening also results in leakage of cytochrome c, which forms a part of the apoptosome that trigger apoptosis. This is evidenced by the migration of PS to the outer leaflet as detected by Annexin V. The depleted ATP levels and increased ROS produce a vicious cycle leading to progressive mitochondrial dysfunction and cell necrosis.[22-24]

In conclusion, this study reveals the dramatic impairment of mitochondrial function by TPL. The effects of TPL occur at clinically relevant exposures. The effects on mitochondrial function therefore are likely to play a role in the biological mechanisms responsible for therapeutic benefit of TPL.

5. REFERENCES

1. S. M. Kupchan, W. A. Court, R. G. Dailey Jr., C. J. Gilmore, R. F. Bryan, Triptolide and tripdiolide, novel antileukemic diterpenoid triepoxides from Tripterygium wilfordii, J. Am. Chem. Soc. 94(20), 7194-7195 (1972).
2. S. Yang, J. Chen, Z. Guo, X. M. Xu, L. Wang, X. F. Pei, J. Yang, C. B. Underhill, L. Zhang, Triptolide inhibits the growth and metastasis of solid tumors, Mol. Cancer Ther. 2(1), 65-72 (2003).
3. T. Tengchaisri, R. Chawengkirttikul, N. Rachaphaew, V. Reutrakul, R. Sangsuwan, S. Sirisinha, Antitumor activity of triptolide against cholangiocarcinoma growth in vitro and in hamsters, Cancer Lett. 133(2), 169-175 (1998).
4. Y. S. Wei, I. Adachi, Inhibitory effect of triptolide on colony formation of breast and stomach cancer cell lines, Chung Kuo. Yao. Li. Hsueh. Pao. 12(5), 406-410 (1991).
5. X. Tao, P. E. Lipsky, The Chinese anti-inflammatory and immunosuppressive herbal remedy *Tripterygium wilfordii* Hook. f., Rheum. Dis. Clin. North Amer. 26, 29-50 (2000).
6. X. Tao, J. J. Cush, M. Garret, P. E. Lipsky, A Phase I study of ethyl acetate extract of the Chinese antirheumatic herb *Tripterygium wilfordii* Hook. f. in rheumatoid arthritis. J. Rheumatol. 28, 2160-2167 (2001).
7. X. C. Tao, V. Pinyopornpanit, J. Younger, J. J. Cai, Y. Mao, L. Ma, S. Chang, P. E. Lipsky, A Phase I/II double-blind, placebo-controlled trial of an extract of *Tripterygium wilfordii* Hook f. shows benefit in patients with rheumatoid arthritis. Abstract, American College of Rheumatology, 65th Ann. Mtg., San Francisco, CA., Nov. 11-17, 2001. Arthritis Rheum. 44(suppl.).
8. J. L. Faul, T. Nishimura, G. J. Berry, G. V. Benson, R. G. Pearl, P. N. Kao, Triptolide attenuates pulmonary arterial hypertension and neointimal formation in rats, Am. J. Respir. Crit. Care Med. 162(6), 2252-2258 (2000).
9. G. Krishna, K. Liu, H. Shigemitsu, M. Gao, T. A. Raffin, G. D. Rosen, PG490-88, a derivative of triptolide, blocks bleomycin-induced lung fibrosis, Am. J. Pathol. 158(3), 997-1004 (2001).
10. E. W. Chan, S. C. Cheng, F. W. Sin, Y. Xie, Triptolide induced cytotoxic effects on human promyelocytic leukemia, T cell lymphoma and human hepatocellular carcinoma cell lines, Toxicol. Lett. 122(1), 81-87 (2001).
11. H. Liu, Z. H. Liu, Z. H. Chen, J. W. Yang, L. S. Li, Triptolide: a potent inhibitor of NF-kappa B in T-lymphocytes. Acta Pharmacol. Sin. 21(9), 782-786 (2000).
12. G. Zhao, L. T. Vaszar, D. Qiu, L. Shi, P. N. Kao, Anti-inflammatory effects of triptolide in human bronchial epithelial cells, Am. J. Physiol. Lung Cell Mol. Physiol. 279(5), L958-L966 (2000).

13. B. J. Chen, C. Liu, X. Cui, J. M. Fidler, N. J. Chao, Prevention of graft-versus-host disease by a novel immunosuppressant, PG490-88, through inhibition of alloreactive T cell expansion, Transplantation 70 (10), 1442-1447 (2000).
14. D. Qiu, G. Zhao, Y. Aoki, L. Shi, A. Uyei, S. Nazarian, J. C. Ng, P. N. Kao, Immunosuppressant PG490 (triptolide) inhibits T-cell interleukin-2 expression at the level of purine-box/nuclear factor of activated T-cells and NF-kappaB transcriptional activation, J. Biol. Chem. 274(19), 13443-13450 (1999).
15. N. Dias, C. Bailly, Drugs targeting mitochondrial functions to control tumor cell growth, Biochem. Pharmacol. 70(1), 1-12 (2005).
16. R. Kim, Recent advances in understanding the cell death pathways activated by anticancer therapy, Cancer 103(8), 1551-1560 (2005).
17. T. Asakura, K. Ohkawa, Chemotherapeutic agents that induce mitochondrial apoptosis, Curr. Cancer Drug Targets 4(7), 577-590 (2004).
18. T. Tengchaisri, R. Chawengkirttikul, N. Rachaphaew, V. Reutrakul, R. Sangsuwan, S. Sirisinha, Antitumor activity of triptolide against cholangiocarcinoma growth in vitro and in hamsters, Cancer Lett. 133(2), 169-175 (1998).
19. Y. S. Wei, I. Adachi, Inhibitory effect of triptolide on colony formation of breast and stomach cancer cell lines, Chung Kuo. Yao. Li. Hsueh. Pao. 12(5), 406-410 (1991).
20. J. Y. Xu, J. Yang, L. Z. Li, Antitumor effect of Tripterygium wilfordii, Chung Kuo. Chung His. I. Chieh Ho. Tsa. Chih. 12(3), 161 (1992).
21. W. Z. Gu, S. R. Brandwein, Inhibition of type II collagen-induced arthritis in rats by triptolide, Int. J. Immunopharmacol. 20(8), 389-400 (1998).
22. The Invitrogen online catalog, accessed 6/18/05. http://probes.invitrogen.com/media/publications/508.pdf
23. V. Petronilli, G. Miotto, M. Canton, M. Brini, R. Colonna, P. Bernardi, F. Di Lisa, Transient and long-lasting openings of the mitochondrial permeability transition pore can be monitored directly in intact cells by changes in mitochondrial calcein fluorescence, Biophys. J. 76(2), 725-734 (1999).
24. V. Petronilli, G. Miotto, M. Canton, R. Colonna, P. Bernardi, F. Di Lisa, Imaging the mitochondrial permeability transition pore in intact cells, Biofactors 8(3-4), 263-272 (1998).

IMMUNOHISTOCHEMICAL IDENTIFICATION AND LOCALIZATION OF ENDOGENOUS ENDOSTATIN AND ITS RELATED PEPTIDES IN MURINE TUMORS

Jianzhong Sun, Ivan Ding, Bruce Fenton, Won Sam Yi, and Paul Okunieff [*]

Abstract: Endostatin, a fragment of the C-terminal domain of mouse collagen XVIII, is a recently demonstrated endogenous inhibitor of tumor angiogenesis. Although endostatin can be detected in blood and urine of tumor-bearing as well as normal mice, the exact localization of the endogenous protein and its related peptides in tumor tissues is unknown. We used immunohistochemistry and immunoblotting to identify endostatin tissue location and staining patterns in tumor, as well as to determine the differences in the levels of endostatin expression between tumor cells (in vitro) and tumor tissues (in vivo). Using a specific polyclonal antibody against murine endostatin, we quantitatively determined the levels of endostatin in five murine mammary tumors and the KHT sarcoma by Western blotting. The staining patterns for this protein in tumor sections were examined histologically by immunohistochemistry. Our results show that: 1) Endogenous endostatin and its related peptides are widely distributed in all in vivo tumor types tested, but not in most of the cultured tumor cell lines. 2) Endogenous endostatin stained most tumor stromal components, including vessel walls, basement membranes, extracellular spaces, and tumor cells. 3) Staining patterns and localization of endostatin and thrombospondin-1 were similar in these tumor sections.

1. INTRODUCTION

Angiogenesis, the formation of new blood vessels, is essential for both physiological and pathological conditions in many diseases,[1-3] including tumor growth.[4-7] Endostatin protein is a proteolytic polypeptide derived from its parent protein, collagen XVIII,[8, 9] through the cleavage of an Ala-His linkage by both a metal-dependent enzyme and

[*]Department of Radiation Oncology, University of Rochester School of Medicine, Rochester, NY 14642 USA

elastase.[10] In vitro studies have shown that endostatin specifically inhibits endothelial proliferation, without a direct effect on tumor or other non-neoplastic cell growth.[11-14] While endostatin inhibits tumor angiogenesis and reduces tumor growth, the precise in vivo mechanism of its action is unclear.[15] In mice, administration of endostatin protein decreased both primary and secondary tumors.[8, 16] The endostatin accumulates in vessel walls and basement membranes of both the developing embryo and the adult mouse.[9] Chang et al. showed that endostatin binds with the basement membrane and endothelial cells of most normal and tumor tissues.[17] Although the biological implications of endostatin binding sites in normal and tumor tissues are unclear, accumulation in vessel walls makes this protein ideally suited to regulate tissue angiogenesis. In this study, we determined the levels of endogenous endostatin in six murine primary tumor types by Western blot, and defined immunohistochemical staining patterns for endostatin and its related peptides. Finally, the localization of another anti-angiogenic growth factor, thrombospondin-1, was examined and compared to that of endostatin.

2. METHODS

2.1. Murine Tumor and Tissue Preparation

Six isotransplanted murine tumors were used: four established mammary tumors (MCa-4, MCa-29, MCa-K, and MCa-35), a spontaneous mammary tumor (Sp-BSB) in C3H/Hej retired breeder mice, an early generation of Sp-BSB (SP-BSB1), and KHT sarcomas. Single tumor cells were inoculated i.m. into the hind limbs of 6-8 week-old female C3H/Hej mice (NCI/NIH, Frederick, MD). Tumors were used at volumes of 200-300 mm^3. One portion of the tumors was used for frozen and paraffin slides, and another portion was snap-frozen in liquid nitrogen and stored at –80°C until use.

2.2. Immunohistochemistry

The 5 μm sections were cut and fixed in pre-cooled acetone for 10 minutes. Frozen sections were quenched in 3% hydrogen peroxide, blocked for 30 min in 5% normal rabbit serum, and incubated at 4°C overnight in a 1:100 dilution of either endostatin (R&D Systems) or thromobospondin-1 (NewMarkers, Union City, CA) antibodies. Subsequent detection was performed using Ellits ABC kit (Vector Lab., CA).

2.3. Western Blotting

Endostatin-transfected MCa-35 cells were grown to 80% confluence, and protein lysate was prepared as a positive control (P. control). 50 μg tumor lysate was electrophoresed in a 7.5% polyacrylamide slab gel and then transferred to a polyvinylidene difluoride membrane (Millipore). The membrane was blocked for 1 h with 10% nonfat dried milk in TBS, and then incubated overnight at 4°C in a 1:500 dilution of respective antibodies in TBS buffer. After three washes with TBS containing 0.1% Tween-20, the membrane was incubated for 45 min at room temperature in a 1:4000 dilution of anti-rabbit secondary antibody linked with horseradish peroxidase (Santa Cruz). Detection was carried out with the ECL kit (Amersham).

3. RESULTS AND DISCUSSION

Endogenous endostatin protein and or fragments of the parent protein were detectable in all tumors tested. Using immunohistochemistry, the localization of endostatin protein was identified and categorized into four staining patterns: basement membrane, vessel wall, extracellular matrix, and cellular staining, as shown in Figure 1 for MCa-4 tumors. Staining patterns for both endostatin and thrombospondin-1 were similar, and high levels of these proteins were observed in several tumor tissue compartments (Figures 1-3). Within the basement membrane, endostatin staining was observed in three locations: 1) around tumor capsules (Figure 1d), 2) around tumor nests (Figure 1e), and 3) around defective basal membranes (Figure 1f).

Endostatin immunoreactivity was also detected in the vessel walls, including staining of large size vessels (Figure 1j); tumor stromal vessels (Figure 1k); and microvessels of intertumoral nests (Figure 1l). The third staining pattern is the extracellular matrix components inside tumor stroma (Figure 1a), including perivascular spaces (Figure 1b) and the spaces between tumor cells (Figure 1c). In the cultured tumor cells, only the very slow growing MCa-29 cell line expressed endostatin or its related peptides (Figure 3b). In vivo, however, many of the mammary tumors demonstrated tumor cellular staining for endostatin. This staining was either diffuse (Figure 1g) or focal (Figure 1h), and staining of the tumor cell membrane was also sometimes observed (Figure 1i).

Sasake et al. have shown that endostatin is widely localized in the vessel walls and basement membranes by immunofluorescence techniques.[9] They also demonstrated that endostatin and its related proteolytic peptides exist in most mouse tissue extracts, as well as in human serum and urine.[11] Using radioimmunoassays, they quantitatively determined the amount of endostatin in both blood and local tissues and concluded that most normal organs produced endostatin and/or related fragments locally. However, only endostatin is a major active component in the blood.[11] The production of endostatin and its related fragments from parental protein C18 is similar in both normal and tumor tissues.[11] However, different organs may generate different sizes of endostatin-related proteins due to different proteolytic enzyme systems. Therefore, tumors expressing endostatin may not produce the effective C18 fragment length/endostatin and instead may produce longer fragment length/endostatin related peptides. Specific enzymes for each proteolytic digestion process remain to be identified.

Tumor tissue extracts have detectable and varied protein levels for endostatin and its related fragments as detected by Western blot (Figure 4). However, different sizes of endostatin and its related peptides were observed depending on tumor types. All tumor tissues tested had detectable endogenous endostatin-related fragments, which included major 25kDa and other large molecular weight fragments. For example, MCa-4, KHT and BSB1 had a second larger molecular weight band (27kDa) in addition to the major 25kDa protein. MCa-4 and KHT tumors had third bands around 39kDa and 32kDa, respectively. In the cultured murine normal and tumor cells, only MCa-29 cells had substantial endogenous endostatin, while highly passaged mammary carcinoma (EMT6) and macrophage (RAW2616) cell lines had a weakly detectable 25 kDa major band.

Recently Sasaki et al. also reported that endostatin was detected in most cultured human and murine epithelial, fibroblast cell lines, but not in tumor lines.[11] Several possible explanations exist for the presence of endostatin and its related peptides in

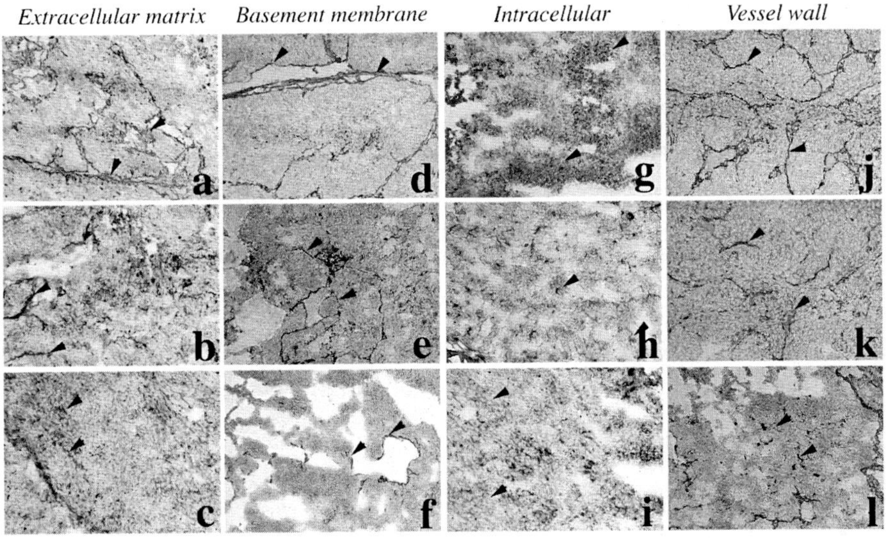

Figure 1. Immunohistochemical determination of endostatin and its related peptides in MCa-4 tumors.

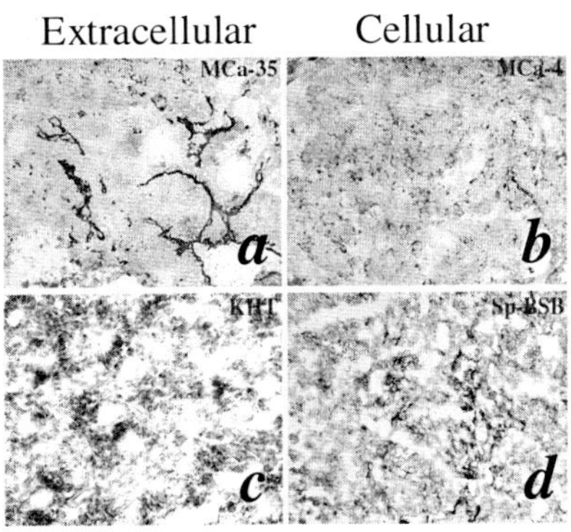

Figure 2. Immunohistochemical determination of thrombospondin protein in four tumor types. All four staining patterns were observed in these tumor tissues.

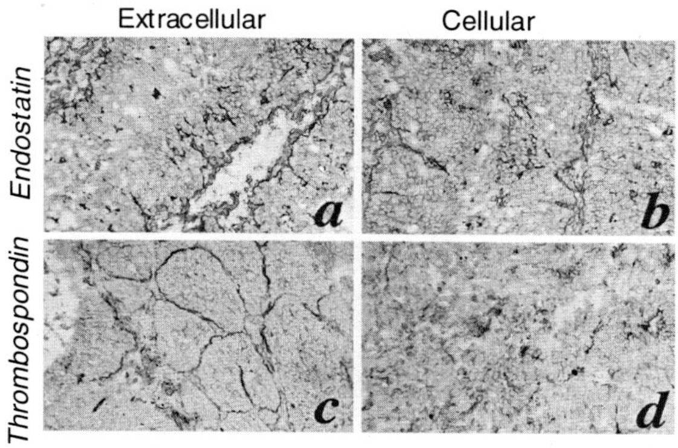

Figure 3. Immunohistochemical determination of endostatin and thrombospondin-1 proteins in MCa-4 tumors. Staining for endostatin (a, b). Staining for thrombospondin-1 (c, d).

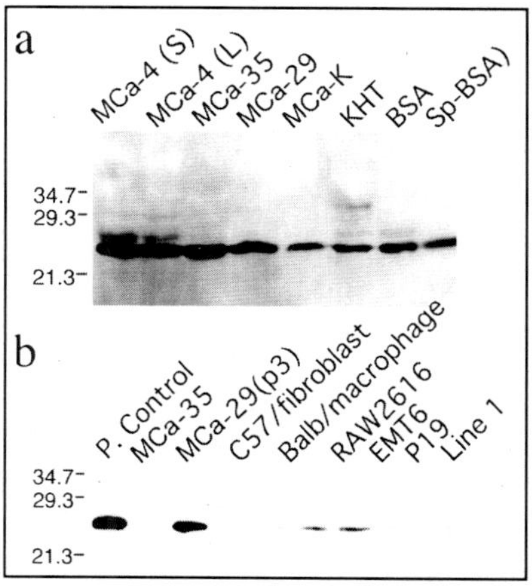

Figure 4. Western blot analysis for endostatin proteins in cell lysates from tumors (a) or cultured cells (b) are shown. Tumors include those from small (S) and large (L) MCa-4 murine mammary adenocarcinomas. There was no clear evidence that tumors of increased aggressiveness expressed decreased levels of endostatin. Likewise, endostatin was invariably expressed by tumor tissues and commonly expressed in tumor cells in-vitro. Endostatin like protein were common in tumor as evidenced by additional lines on the Western blot (e.g. MCa-4 and KHT).

tumors and their microenvironment: 1) Circulating endostatin may leak out of defective vessels, which may account for most of the intratumoral stroma staining. 2) Local endothelial cells may produce endostatin, which either localizes to the extracellular matrix or exists as a component of the vessel walls. 3) Least likely, tumor cells themselves, particularly undifferentiated tumor cells, may produce small amounts of endostatin and its related peptides. Endostatin tissue localization detected by immunohistochemistry strongly suggests that endostatin and its related peptides may interact with other structural extracellular molecules. These postulated interactions require additional biochemical characterization. In this study, several tumor types with different malignant potential were used in order to examine the relationship between endostatin expression levels and tumor malignant potential. For instance, MCa-29, MCa-35 and MCa-K have a high incidence of lung metastasis in comparison to MCa-4 and Sp-BSB tumors. However, endostatin protein expression levels detected by immunohistochemistry or Western blot analysis were not associated with tumor metastatic potential in this group of tumors.

Finally, we found a similar tumor tissue distribution of the two antiangiogenic growth factors, endostatin and thrombospondin-1 proteins, by immunohistochemistry. As shown in Figure 2, the staining patterns of thrombospondin were observed using anti-thrombospondin-1 antibody in the tumor types we tested. Thrombospondin staining could be observed as membrane or vessel staining (Figure 2a) or as cellular staining (Figure 2b-2d). Co-localization between endostatin and thrombospondin immunoreactivity in tumor tissues was frequently noted. Extracellular staining for endostatin (Figure 3a) and thrombospondin (Figure 3c), and cellular staining for endostatin (Figure 3b) and thrombospondin (Figure 3d), were particularly prominent in the MCa-4 tumor. This implies that tumor antiangiogenesis may require co-expression of more than one antiangiogenic molecule, and different antiangiogenic factors may combat different angiogenic components.

Using three types of human liver tissue (normal, cirrhosis, and cancer) for a C18 in situ hybridization, Schuppan et al. found that hepatocytes are a major source of C18.[18] They found that C18 was highest in cancerous hepatocytes but lowest in normal hepatocytes. Some C18 mRNA expression was also found in bile duct epithelial and endothelial cells, but not in fibroblasts and monocytes. They suggested that: 1) the liver regulates C18 differently from other collagens due to the C-terminus fragment endostatin, 2) monomeric endostatin may serve as a competitive antagonist for the trimeric basement membrane C18, binding the endostatin receptor on endothelial cells, and 3) endostatin may sequester a growth factor for endothelial cell growth. Teleologically, the presence of endostatin in the blood vessels of normal tissues is explainable. Cellular turnover in most vasculature is slow. In certain environments, however, such as in granulation tissue, vascular proliferation occurs at a pace that exceeds even the vast majority of tumor tissue. The vascular localization of endostatin would therefore be critical in preventing uncontrolled proliferation in normal tissues. The endostatin we detected in tumors, however, is enigmatic. The fragment size distribution of endostatin was not particularly abnormal and its levels were not reduced in tumor tissues compared to normal tissues. Hence, either the levels of endostatin and/or its related fragments in tumors were insufficient to prevent tumor angiogenesis and growth, or endostatin and/or its related peptides were somehow unable to exert their effect (perhaps due to specific stromal binding or inhibitory factors produced by tumors). Schuppan and colleagues found that C18 was most highly expressed in the highly angiogenic hepatocellular carcinomas, lowest in normal hepatocytes, and intermediate in cirrhosis.[18] The possible explanations for these findings are: (1) endostatin is being produced in a homeostatic effort to combat angiogenesis, or (2) the abnormal lengths of endostatin fragments in tumor are actually improving tumor angiogenesis and growth. The latter explanation deserves further study, since it may lead to a cancer treatment by cleavage of endogenous endostatin related fragments to the effective endostatin length.

Endostatin or its related peptides are seen in many epithelial and mesenchymal tumors grown in mice. Protein expression for endostatin or its related fragments is present not only in tumor blood vessels and tumor stroma, but also within tumor cells. No correlation was found between endostatin staining and tumor aggressiveness in these four well-characterized mammary tumors. In contrast, angiogenesis of murine tumors typically parallels aggressiveness.[19] Some of the endostatin fragments are larger than the naturally occurring endostatin, suggesting that different fragment lengths may have different activities or effectiveness. An explanation for these observations includes the possibility that some longer fragments of endostatin may actually be angiogenic.

4. REFERENCES

1. J. C. Bout, F. Soncin, B. Vandenbunder, Therapeutic prospects of angiogenesis inhibitors in cancerology, Rev. Mal. Respir. 16, 854-855 (1999).
2. E. Keshet, S. A. Ben-Sasson, Anticancer drug targets: approaching angiogenesis, J. Clin. Invest. 104, 1497-1501 (1999).
3. Y. Cao, Therapeutic potentials of angiostatin in the treatment of cancer, Haematologica 84, 643-650 (1999).
4. J. Folkman, Angiogenesis in cancer, vascular, rheumatoid and other disease, Nat. Med. 1, 27-31 (1995).
5. J. Folkman, Angiogenesis inhibitors generated by tumors, Mol. Med. 1, 120-122 (1995).
6. P. Hahnfeldt, D. Panigrahy, J. Folkman, L. Hlatky, Tumor development under angiogenic signaling: a dynamical theory of tumor growth, treatment response, and postvascular dormancy, Cancer Res. 59, 4770-4775 (1999).
7. A. Le Querrec, D. Duval, G. Tobelem, Tumour angiogenesis, Baillieres Clin. Haematol. 6, 711-730 (1993).
8. M. S. O'Reilly, T. Boehm, Y. Shing, N. Fukai, G. Vasios, W. S. Lane, E. Flynn, J. R. Birkhead, B. R. Olsen, J. Folkman, Endostatin: an endogenous inhibitor of angiogenesis and tumor growth, Cell 88, 277-285 (1997).
9. T. Sasaki, H. Larsson, J. Kreuger, M. Salmivirta, L. Claesson-Welsh, U. Lindahl, E. Hohenester, R. Timpl, Structural basis and potential role of heparin/heparan sulfate binding to the angiogenesis inhibitor endostatin, Embo. J. 18, 6240-6248 (1999).
10. W. Wen, M. A. Moses, D. Wiederschain, J. L. Arbiser, J. Folkman, The generation of endostatin is mediated by elastase, Cancer Res. 59, 6052-6056 (1999).
11. T. Sasaki, N. Fukai, K. Mann, W. Gohring, B. R. Olsen, R. Timpl, Structure, function and tissue forms of the C-terminal globular domain of collagen XVIII containing the angiogenesis inhibitor endostatin, Embo. J. 17, 4249-4256 (1998).
12. M. Dhanabal, R. Ramchandran, M. J. Waterman, H. Lu, B. Knebelmann, M. Segal, V. P. Sukhatme, Endostatin induces endothelial cell apoptosis, J. Biol. Chem. 274, 11721-11726 (1999).
13. M. Dhanabal, R. Ramchandran, R. Volk, I. E. Stillman, M. Lombardo, M. L. Iruela-Arispe, M. Simons, V. P. Sukhatme, Endostatin: yeast production, mutants, and antitumor effect in renal cell carcinoma, Cancer Res. 59 , 189-197 (1999).
14. L. Taddei, P. Chiarugi, L. Brogelli, P. Cirri, L. Magnelli, G. Raugei, M. Ziche, H. J. Granger, V. Chiarugi, G. Ramponi, Inhibitory effect of full-length human endostatin on in vitro angiogenesis, Biochem. Biophys. Res. Commun. 263, 340-345 (1999).
15. P. Blezinger, J. Wang, M. Gondo, A. Quezada, D. Mehrens, M. French, A. Singhal, S. Sullivan, A. Rolland, R. Ralston, W. Min, Systemic inhibition of tumor growth and tumor metastases by intramuscular administration of the endostatin gene, Nat. Biotechnol. 17, 343-348 (1999).
16. T. Boehm, J. Folkman, T. Browder, M. S. O'Reilly, Antiangiogenic therapy of experimental cancer does not induce acquired drug resistance see comments, Nature 390, 404-407 (1997).
17. Z. Chang, A. Choon, A. Friedl, Endostatin binds to blood vessels in situ independent of heparan sulfate and does not compete for fibroblast growth factor-2 binding, Am. J. Pathol. 155, 71-76 (1999).
18. D. Schuppan, T. Cramer, M. Bauer, T. Strefeld, E. G. Hahn, H. Herbst, Hepatocytes as a source of collagen type XVIII endostatin letter, Lancet 352, 879-880 (1998).
19. H. Suit, A. Allam, J. Allalunis-Turner, W. Brock, T. Girinsky, S. Hill, N. Hunter, L. Milas, R. Pearcey, L. Peters, et al. Is tumor cell radiation resistance correlated with metastatic ability? Cancer Res. 54(7):1736-1741 (1994).

IMPACT OF HYPOXIC AND ACIDIC EXTRACELLULAR CONDITIONS ON CYTOTOXICITY OF CHEMOTHERAPEUTIC DRUGS

Oliver Thews[*], Birgit Gassner[**], Debra K. Kelleher[*], Gerald Schwerdt[**], Michael Gekle[**]

Abstract: In the microenvironment of solid growing tumors, pronounced hypoxia or extracellular acidosis is commonly found. The aim of this study was the analysis of the cytotoxic effect of different chemotherapeutic agents (cisplatin, daunorubicin, docetaxel) under these conditions *in vitro*. Prostate carcinoma cells (R3327-AT1) were exposed to hypoxia (pO_2<0.5 mmHg) or extracellular acidosis (pH=6.6) for 6h. After 3h, cytotoxic drugs were added. The cytotoxic effect was assessed by measuring caspase 3-activity (apoptosis), LDH release (necrosis) and repopulation of the cells after chemotherapy (cell death). Compared to aerobic control conditions, severe hypoxia over 6h *per se* led to a slight increase in apoptosis, necrosis and cell death. With all three chemotherapeutic agents, hypoxia led to a reduced (by approx. 25%) caspase 3-activity and a marked increase in necrosis. However, the overall cytotoxicity of the drug was not affected by O_2-deficiency. By contrast, during extracellular acidosis, the cytotoxic effect of daunorubicin was reduced by 40%, preferentially due to a marked reduction in apoptosis. With cisplatin and docetaxel no change in overall cell death was detected. However, for daunorubicin the tumor-pH seems to have a strong impact on cytotoxicity. With this chemotherapeutic drug the therapeutic efficacy is markedly reduced in an acidotic environment.

1. INTRODUCTION

In many experimental and human solid-growing tumors, pronounced oxygen deficiency (pO_2<1 mmHg) and/or extracellular acidosis (pH<6.8) due to obligatory glycolysis is commonly found[1]. Numerous previous studies suggest that hypoxia in solid tumors is one

[*] Institute of Physiology and Pathophysiology, University of Mainz, 55099 Mainz, Germany
[**] Institute of Physiology, University of Wuerzburg, 97070 Wuerzburg, Germany

of the major reasons for a limited efficacy of several non-surgical treatment modalities. Teicher et al.[2] showed that the *in vivo* anti-neoplastic activity of several drugs in experimental sarcomas of mice was correlated to tumor perfusion. The authors assumed that well-perfused tissue areas corresponded to normoxic regions, whereas hypoxia originated from poorly perfused regions. They found that several chemotherapeutic agents induced a more pronounced cell kill in "oxygenated" tumor areas than in "hypoxic" regions indicating a possible O_2-sensitivity of these drugs. On the other hand, extracellular acidosis has also been blamed for a reduction in the cytotoxicity of drugs[3]. Various reasons for the impact of hypoxia or an extracellular acidification on cytotoxicity have been discussed: (1) hypoxia can cause a cell-cycle arrest and in turn reduce the efficacy of agents acting only on proliferating cells[4,5], (2) oxygen might directly influence the mechanism of action (pharmacodynamics) of anti-neoplastic drugs (e.g., doxorubicin by formation of reactive oxygen species)[5,6], (3) hypoxic tissues show a pronounced extracellular acidity[1] which might influence the intra-/extracellular drug distribution and by this lead to a reduced intracellular concentration of chemotherapeutic agents[4,7].

The aim of this study was the analysis of the cytotoxic effect of different chemotherapeutic agents (cisplatin, daunorubicin, docetaxel) under hypoxic and acidic conditions *in vitro*.

2. MATERIAL AND METHODS

2.1. Cell line

The subline AT1 of the R-3327 Dunning-prostate carcinoma of the rat was used in all experiments. Cells were grown in RPMI medium (11.1 mmol/L glucose, 2.05 mmol/L L-glutamine) supplemented with 10% fetal calf serum (FCS) at 37°C under a humidified 5% CO_2 atmosphere and passaged once per week. Twenty-four hours prior to the experiments, cells were transferred to RPMI medium without additional FCS supplementation. For experiments under control and hypoxic conditions, the medium was buffered with 20mM HEPES adjusted to a pH of 7.4. In hypoxia experiments, cells were gassed with a mixture of 95% N_2 and 5% CO_2. For acidic conditions, cells were incubated in a medium buffered with 20 mM MES (morpholinoethanesulfonic acid) + 4.51 mM $NaHCO_3$ resulting in a pH of 6.6 for up to 6h.

2.2. Chemotherapy and cytotoxicity measurements

The impact of the microenvironment on the cytotoxicity of three different chemotherapeutic drugs was analyzed: daunorubicin (DNR), cisplatin, docetaxel. All drugs (Sigma-Aldrich, Taufkirchen, Germany) were dissolved in medium at final concentrations of 10 μM (DNR), 150 μM (cisplatin), and 1 μM (docetaxel). In order to assess the cytotoxic activity of the chemotherapeutic drugs under different conditions, the relative caspase 3-activity normalized with respect to untreated control cells and necrosis formation were measured. In addition, the cellular repopulation 42h after chemotherapy was determined. For these experiments, cells were grown as a confluent monolayer. Three hours before application of the chemotherapeutic drugs, cells were exposed to either acidic conditions (pH=6.6, ambient pO_2), hypoxia (pH=7.4, pO_2<0.5 mmHg) or control conditions (pH=7.4, ambient pO_2). Subsequently, cells were incubated with one

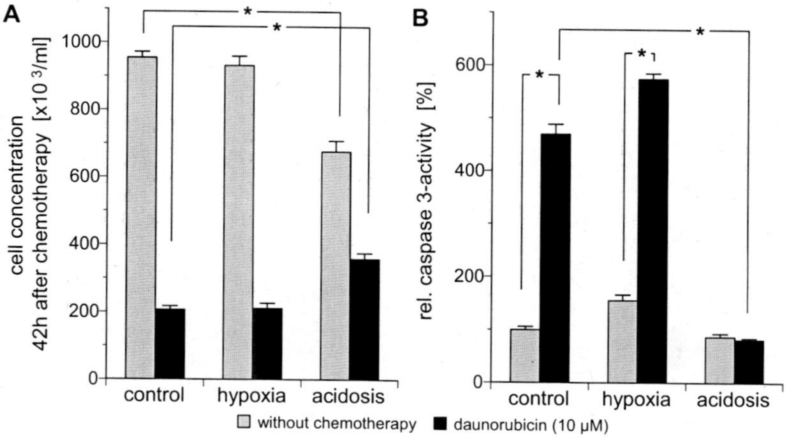

Figure 1. Impact of environmental hypoxia and acidosis on the cytotoxic activity of daunorubicin chemotherapy (10 µM, 3h) measured by (A) tumor cell repopulation over 42h after chemotherapy and (B) the induction of apoptosis. Data are expressed as mean ± SEM of n=5 experiments, (*) $p<0.05$. The cell concentration in the chemotherapy-treated groups was significantly ($p<0.05$) lower than in untreated controls, under all three environmental conditions.

of the chemotherapeutic drugs for 3h at 37°C, whereby the acidic, hypoxic or control conditions were maintained. After the chemotherapy period the medium was again replaced. For caspase 3-activity and necrosis measurements, cells were exposed to serum-free control medium (pH=7.4, ambient pO_2) for a further 3h, after which the parameters were measured as described below. For repopulation experiments, following incubation with the drugs, the medium was replaced with the FCS-containing control medium, and cells were maintained under normal cell culture conditions for 42h. Thereafter, the cell concentration in the Petri dishes was determined using an automatic cell counter (Z2, Coulter, Krefeld, Germany).

Caspase 3-activity was determined as described previously[8] using a caspase 3-activity kit (Clontech Laboratories, Heidelberg, Germany). For this assay, cells were incubated with a lysis buffer (100 µL) for 20 min, harvested and centrifuged. The supernatant (60 µL) was incubated with the caspase 3-substrate DEVD-AFC (40 µM final concentration) at 37°C in a 96-well plate. Fluorescence of the cleaved product AFC was measured at 405 nm excitation and 535 nm emission wavelengths using a fluorescence microplate reader (Victor2, Wallac, Turku, Finlan d). All measurements were normalized with respect to the protein content in each probe.

Necrosis was measured by lactate dehydrogenase (LDH) release from the cells into the medium. Activity of LDH in media and cell lysates was determined with an automatic analyzer (Cobas-Mira, Roche Diagnostics, Mannheim, Germany).

3. RESULTS

Exposing tumor cells to hypoxic environmental conditions without additional chemotherapy has no marked impact on tumor cell proliferation. Keeping cells for 6h

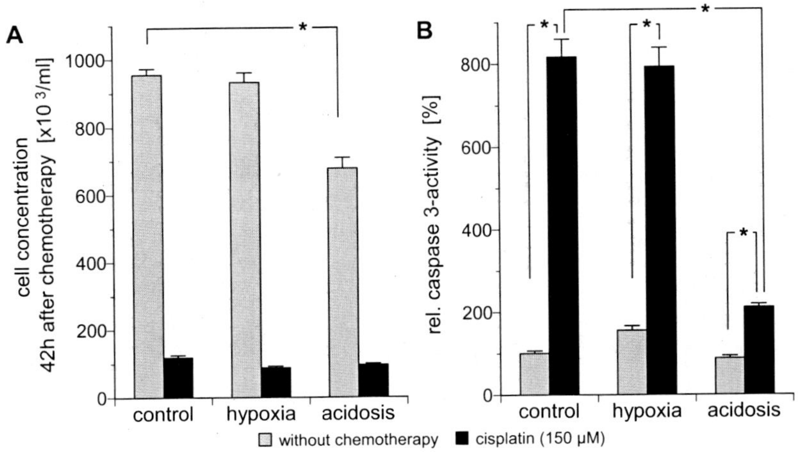

Figure 2. Impact of environmental hypoxia and acidosis on the cytotoxic activity of cisplatin (150 μM, 3h) measured by (A) tumor cell repopulation over 42h after chemotherapy and (B) the induction of apoptosis. Data are expressed as mean ± SEM of n=5 experiments, (*) $p<0.05$. The cell concentration in the chemotherapy-treated groups was significantly ($p<0.05$) lower than in untreated controls, under all three environmental conditions.

under hypoxic conditions led to almost no change in the cell concentration 42h later with only a minor increase in the caspase 3-activity (Fig. 1). The extent of necrosis (measured by the fraction of LDH release into the medium) increased from 6 ± 2% under control conditions to 22 ± 16% under hypoxic conditions. However, this difference was not statistically significant due to large intertumoral variability. Keeping cells under acidic conditions (pH=6.6, without chemotherapy) for 6 h led to a decrease in cell survival of approx. 20% (Fig. 1A). Since apoptosis in the acidosis groups was not significantly different from that in cells under control conditions (Fig. 1B), this reduction in cell growth seems to be preferentially due to an increase in necrosis from 6 ± 2% to 23 ± 11% under acidic conditions ($p=0.06$).

Incubating AT1-cells with any of the three chemotherapeutic drugs under normal cell culture conditions led to a substantial cell kill, which was most pronounced with cisplatin. With this drug, the number of surviving cells was significantly reduced by 85% and apoptosis formation increased by 700% (Fig. 2). With daunorubicin, the number of surviving cells was also markedly reduced indicating a pronounced cell kill by apoptosis (Fig. 1). With docetaxel the number of tumor cells after chemotherapy was halved (compared to non-treated controls, $p<0.05$) (Fig. 3) indicating that this drug at the dose used (1 μM) has the weakest cytotoxic impact.

Exposing the cells to severe **hypoxia** ($pO_2<0.5$ mmHg) during incubation with these drugs had almost no effect on the cytotoxicity. With all three chemotherapeutic agents the number of surviving tumor cells as well as the induction of apoptosis was almost the same as under control conditions ($pO_2 \approx 150$ mmHg) (Figs. 1-3). The only difference was seen in the amount of necrosis formation when cells were incubated with cisplatin. With this drug the fraction of LDH found in the medium increased significantly from 22 ± 16% without chemotherapy to 65 ± 8% after cisplatin treatment under hypoxia. In cells treated with cisplatin under normal cell culture conditions this fraction was 6 ± 2%. However, this marked increase in necrosis seems to have almost no impact on the overall cell

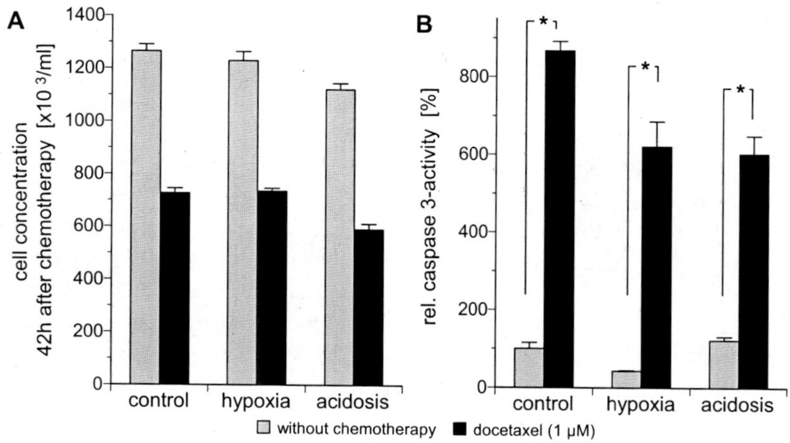

Figure 3. Impact of environmental hypoxia and acidosis on the cytotoxic activity of docetaxel (1 μM, 3h) measured by (A) tumor cell repopulation over 42h after chemotherapy and (B) the induction of apoptosis. Data are expressed as mean±SEM of at least 5 experiments, (*) $p<0.05$. The cell concentration in the chemotherapy-treated groups was significantly ($p<0.05$) lower than in untreated controls, under all three environmental conditions,.

survival since the number of surviving cells after cisplatin incubation under hypoxic or control conditions was almost the same (Fig. 2A).

Incubating the cells with the drugs under *acidic conditions* (pH=6.6) showed different effects depending on the agent used. With docetaxel the cell kill, as well as the induction of apoptosis, was not markedly different from that found under control conditions (Fig. 3). Upon exposure to cisplatin, the number of surviving cells was unaffected by the acidic environment (Fig. 2A). However, apoptosis was significantly reduced as compared to control conditions (Fig. 2B). The cytotoxic effect of daunorubicin showed the most pronounced dependency on the extracellular pH. Under acidic conditions the number of surviving cells after chemotherapy was significantly higher as compared to cells treated at normal pH (Fig. 1A). The results of the caspase 3-activity assessment show the pH-effect even more clearly. When cells were incubated with daunorubicin at low pH almost no induction of apoptosis (as compared to cells under acidic conditions without chemotherapy) was seen (Fig. 1B) clearly indicating a marked reduction in cytotoxicity.

4. DISCUSSION

Pronounced hypoxia (which is commonly found in human tumors) may affect certain aspects of the cellular behavior of tumor cells such as proliferation and apoptotic cell death via a p53-dependent pathway[9]. Even though the cell line used in the present study (R-3327 AT-1) expresses p53[10], hypoxia with a $pO_2 < 0.5$ mmHg for 6h was not able to induce apoptosis or reduce the number of proliferating tumor cells (Fig. 1). This might be explained by the fact that the energy supply of the cells was maintained. The glucose concentration in the medium, even after 6h of hypoxia, was 5.4 ± 0.7 mM (data not shown). Extracellular acidosis has also been described as inhibitory for tumor cell proliferation also via a p53-dependent pathway[11]. These authors described a reduced tumor cell proliferation at a pH of 6.6 comparable to that found in the present study (Fig. 1A). Our results however, indicate that the reduced number of surviving cells after exposure to acidic conditions seems to be the result of increased necrosis rather than forced apoptosis (Fig. 1B).

It has been postulated that the adverse microenvironmental conditions (hypoxia, acidosis) in tumors affect the cytotoxic efficacy of chemotherapeutic drugs. In cell culture experiments, studies have demonstrated various chemotherapeutic agents to be more effective in the presence of oxygen[12,13]. In these studies, severe hypoxia was found to reduce the cytotoxicity of doxorubicin or cisplatin, although such a reduction was not seen in the present study. However, the main difference between these studies was that in the present experiments cells were exposed to hypoxia for only 6h whereas in other studies cells were exposed to hypoxia for 24h. On the basis of the results presented here, the possibility of a decrease in the resistance of these cells to the drugs assessed upon incubation under hypoxic conditions for longer periods of time cannot be ruled out. It should be noted however, that when R-3327 AT-1 cells were incubated at $pO_2 < 0.5$ mmHg for 24h, most of the cells already showed signs of dying (even without chemotherapy) as indicated by a detachment of the cells from the Petri dish (data not shown).

Acidic environmental conditions have also been blamed for a reduction in the cytotoxic effect of drugs. Reichert et al.[11] found an increased chemoresistance at a pH of 6.6 with several drugs such as cisplatin, topotecan and vincristine which is in good accordance with the result of the present study (Figs. 1 and 2). The authors, however, also showed these pH-effects to be drug-dependent. For instance, with lomustine they found an increase in cytotoxicity at low pH, which might explain the lack of a pH impact with docetaxel seen in the present study.

In the present study, the overall cell kill with cisplatin was independent of the extracellular pH (Fig. 2A). Previous studies[14,15] demonstrated that with decreasing pH, cisplatin-induced cytotoxicity was markedly increased. One possible explanation for the discrepancy between these results and the present study may be differences in the sensitivities of the cell lines to cisplatin. A further reason could be the high cisplatin dose used in the present study which may induce an almost maximal cytotoxic effect which can not be further amplified by a low pH.

There may be a number of different reasons for the pH-dependency of cytotoxicity. Under the assumption that the uptake of drugs into the cells occurs passively by diffusion of the non-ionized forms of the drugs, the extracellular pH might play an important role should the drug itself act as a weak acid or base. Since daunorubicin is weakly basic the uptake of this drug in an acidic environment is hindered[7]. The uneven distribution of daunorubicin due to the reduced extracellular pH which results in a lower intracellular drug concentration may be responsible for the less cytotoxic effect under these conditions.

On the other hand, the intracellular drug concentration results from the steady state of drug uptake and efflux via active drug transporters (e.g. p-glycoprotein pGP). Since R-3327 AT-1 cells functionally express pGP it is possible that under acidic conditions the activity of this pump is increased. When an inhibitor of pGP (verapamil) was added to the incubation medium of cells being treated with daunorubicin under acidic pH conditions, the cytotoxicity obtained reached that found at normal pH[16]. These results reveal that a pH-dependency of drug efflux pumps seems to be of importance for the chemosensitivity to some drugs.

5. ACKNOWLEDGEMENT

This study was supported by the Deutsche Krebshilfe (grants 106774 and 106906).

6. REFERENCES

1. P. Vaupel, F. Kallinowski, and P. Okunieff. Blood flow, oxygen and nutrient supply, and metabolic microenvironment of human tumors: a review. *Cancer Res.* **49**, 6449-6465 (1989).
2. B.A. Teicher, S.A. Holden, A. al Achi, and T.S. Herman. Classification of antineoplastic treatments by their differential toxicity toward putative oxygenated and hypoxic tumor subpopulations in vivo in the FSaIIC murine fibrosarcoma. *Cancer Res.* **50**, 3339-3344 (1990).
3. N. Raghunand, X. He, van Sluis R., B. Mahoney, B. Baggett, C.W. Taylor, G. Paine-Murrieta, D. Roe, Z.M. Bhujwalla, and R.J. Gillies. Enhancement of chemotherapy by manipulation of tumour pH. *Br. J. Cancer* **80**, 1005-1011 (1999).
4. D.J. Chaplin, M.R. Horsman, M.J. Trotter, and D.W. Siemann. Therapeutic significance of micro-environmental factors, in: *Blood Perfusion and Microenvironment of Human Tumors*, edited by M. Molls and P. Vaupel, (Springer, Berlin, 1998), pp. 131-143.
5. B.A. Teicher. Hypoxia and drug resistance. *Cancer Metastasis Rev.* **13**, 139-168 (1994).
6. P.E. Ganey, F.C. Kauffman, and R.G. Thurman. Oxygen-dependent hepatotoxicity due to doxorubicin: role of reducing equivalent supply in perfused rat liver. *Mol. Pharmacol.* **34**, 695-701 (1988).
7. N. Raghunand, B.P. Mahoney, and R.J. Gillies. Tumor acidity, ion trapping and chemotherapeutics. II. pH-dependent partition coefficients predict importance of ion trapping on pharmacokinetics of weakly basic chemotherapeutic agents. *Biochem. Pharmacol.* **66**, 1219-1229 (2003).
8. G. Schwerdt, R. Freudinger, C. Schuster, S. Silbernagl, and M. Gekle. Inhibition of mitochondria prevents cell death in kidney epithelial cells by intra- and extracellular acidification. *Kidney Int.* **63**, 1725-1735 (2003).
9. M.S. Soengas, R.M. Alarcon, H. Yoshida, A.J. Giaccia, R. Hakem, T.W. Mak, and S.W. Lowe. Apaf-1 and caspase-9 in p53-dependent apoptosis and tumor inhibition. *Science* **284**, 156-159 (1999).
10. D.B. Cooke, V.E. Quarmby, D.D. Mickey, J.T. Isaacs, and F.S. French. Oncogene expression in prostate cancer: Dunning R3327 rat dorsal prostatic adenocarcinoma system. *Prostate* **13**, 263-272 (1988).
11. M. Reichert, J.P. Steinbach, P. Supra, and M. Weller. Modulation of growth and radiochemosensitivity of human malignant glioma cells by acidosis. *Cancer* **95**, 1113-1119 (2002).
12. B.C. Liang. Effects of hypoxia on drug resistance phenotype and genotype in human glioma cell lines. *J. Neurooncol.* **29**, 149-155 (1996).
13. K. Sanna and E.K. Rofstad. Hypoxia-induced resistance to doxorubicin and methotrexate in human melanoma cell lines in vitro. *Int. J. Cancer* **58**, 258-262 (1994).
14. A. Atema, K.J. Buurman, E. Noteboom, and L.A. Smets. Potentiation of DNA-adduct formation and cytotoxicity of platinum-containing drugs by low pH. *Int. J. Cancer* **54**, 166-172 (1993).
15. T.S. Herman, B.A. Teicher, and L.S. Collins. Effect of hypoxia and acidosis on the cytotoxicity of four platinum complexes at normal and hyperthermic temperatures. *Cancer Res.* **48**, 2342-2347 (1988).
16. O. Thews, B. Gassner, D.K. Kelleher, G. Schwerdt, and M. Gekle. Impact of extracellular acidity on the activity of p-glycoprotein and the cytotoxicity of chemotherapeutic drugs. *Neoplasia* **8**, 143-152 (2006).

BRIEF EXPOSURE TO –2 G_Z REDUCES CEREBRAL OXYGENATION IN RESPONSE TO STAND TEST

Cong C.D. Tran, Muriel Berthelot, Xavier Etienne, Pascal Van Beers, Caroline Dussault, Jean-Claude Jouanin [*]

Abstract: The aim of the present experiment was to determine whether a single 30 s of exposure to –2 G_z (foot-to-head inertial forces) as orthostatic stress results in altered brain oxygenation control in response to active standing. Cerebral oxygenation (oxy-Hb), cerebral blood volume (CBV), and mean arterial blood pressure at brain level (MAP_{brain}) were recorded in 12 subjects in supine and then in standing position (10 min), before and after –2 G_z centrifugation. The decrease in oxy-Hb (-5±9 vs –9±10 μM, $P<0.001$) and in CBV (-2±11 vs –4±12 μM, $P<0.05$) upon standing was more important after –2 G_z centrifugation, with unchanged MAP_{brain} (-6±7 vs –6±9 mmHg). These findings suggest a downward shift in the static cerebral autoregulation curve.

1. INTRODUCTION

Orthostatic stress applied in humans allows studying the adaptation mechanisms involved in cerebral perfusion and in brain tissue oxygenation control in the face of potential syncope. Acute exposures to cardiovascular stresses such as head-up tilt (HUT) and lower body negative pressure (LBNP) are commonly used. It is well established that HUT may result in a downward shift1 in the cerebral autoregulatory curve, whereas LBNP may result in a rightward shift.2,3 The shift in the autoregulatory curve has been attributed to the occurrence of cerebral vasoconstriction. Recently, +Gz centrifugation (head-to-foot inertial forces) has been used as cardiovascular stress to find better insights in the mechanisms involved in these shifts of the cerebral autoregulatory curve.4,5 Indeed, +Gz centrifugation may subsequently result in altered cerebrovascular control and orthostatic tolerance due to the reduced cerebral perfusion pressure during +Gz exposure. Cerebral vasodilatation appears during +G_z exposure in response to the decreased cerebral perfusion. Inversely, cerebral vasoconstriction occurs during –G_z exposure (foot-to-head inertial forces) to prevent overperfusion.

The aim of the present work was to determine whether a single exposure of very short duration to low –G_z centrifugation subsequently results in altered brain oxygenation control in response to active standing. We hypothesized that cerebral vasoconstriction

[*] Cong C.D. Tran, Muriel Berthelot, Xavier Etienne, Pascal Van Beers, Caroline Dussault, Jean-Claude Jouanin. Institut de médecine aérospatiale du service de santé des armées, BP 73, 91223 Brétigny/Orge, France.

occurring under $-G_z$ acceleration may result in a downward or a rightward shift in the autoregulatory curve during the immediate recovery period.

2. METHODS

2.1. Subjects

Twelve healthy male non-pilot volunteers participated in this study. The subjects averaged 35 yr in age (range = 25-47 yr), 75 ± 12 kg in weight, and 178 ± 5 cm in height. They were informed of the purpose of the experiment, and the study was carried out with the approval of the national Ethics Committee.

2.2. Physiological measurements

2.2.1. Cerebral oxygenation

Cerebral oxygenation was measured with near infrared spectroscopy (NIRS) which allows to measure continuously and non-invasively changes in brain oxygenation concentration and in cerebral hemodynamics.[6,7] Our study was performed using the NIRO-300 monitor (Hamamatsu Photonics, Hamamatsu, Japan) which produces four wavelengths of near-infrared light (775, 810, 850, and 910 nm) allowing monitoring changes in oxygenated (oxy-Hb) and deoxygenated (deoxy-Hb) hemoglobin concentrations, and cerebral blood volume as changes in total hemoglobin (total-Hb = oxy-Hb + deoxy-Hb) with a modified Beer-Lambert equation.[8] By using a self-adhesive pad, the optodes (light source and light detector) were placed 4.0 cm apart on the left forehead, avoiding the temporal muscle regions. The differential path length factor was set at the level recommended for the adult head[9] to quantify the NIRS data in μM, and the sample frequency was set at 2 Hz.

2.2.2. Beat-to-beat arterial blood pressure

Beat-to-beat arterial blood pressure (BP) was monitored non-invasively using finger plethysmographic techniques (Finapres 2300, Ohmeda, Englewood, CO, USA). Mean arterial pressure (MAP) was one-third of pulse pressure plus the diastolic blood pressure. The BP finger cuff was placed on the left hand and fixed at heart level with an arm sling. MAP at the level of the brain (MAP_{brain}) was calculated by subtracting the appropriate hydrostatic gradient from measurements taken from the finger at heart level.

2.2.3. Recordings

The analogue NIRS signals, BP, and electro-cardiogram (ECG) were sampled simultaneously at 500 Hz per channel and stored by using a digital recorder (MP 100, Biopac, CA, USA). Integrated Acknowledge software allowed off-line data analysis after re-sampling at 50 Hz.

2.3. Protocol

Each subject was exposed to 30 s of $-2\ G_z$ centrifugation. The subject wore no anti-G suit and was asked to be relaxed. To this end the subject underwent one training session in performing the $-2\ G_z$ centrifugation. All subjects were also submitted to stand test-training session to familiarize them with the manoeuvres. Stand test data were collected during identical pre- and post-centrifugation sessions. The pre-centrifugation stand test occurred at the beginning of the morning, and the post-centrifugation one started within 15 min after exiting the gondola. The sequential activities of test subjects were as follows, both before and immediately after centrifugation: 1/ bioinstrumentation in the physiological room, adjacent to the centrifugation testing facility; 2/ 15 min of supine rest followed by 10 min of additional rest for supine control recordings; 3/ active standing in about 5 s and maintained during 10 min. Subjects were instructed to stand without any movement and to breathe with a metronome at 12 breaths.min^{-1}. Before, during and after standing, NIRS variables, BP and ECG were continuously recorded.

2.4. Data analysis

Cerebral autoregulation was assessed by cerebral NIRS variables and MAP_{brain} analysis, whereas autonomic nervous system activity was assessed by arterial baroreflex analysis. For each posture, the values were averaged as the following: 5 min during the supine baseline period prior to active standing, and 5 min during the active standing (from min 2 to min 7). The baroreflex response was analysed in terms of spontaneous baroreflex sensitivity (SBS) at steady state during supine and upright posture. A spontaneous baroreflex sequence[10] was defined by the occurrence of at least three consecutive cardiovascular cycles, in which systolic pressure (minimal change of 1 mmHg) and the following R-R interval (minimal change of 5 ms) either both increased or both decreased. Sequences were only accepted for analysis if the correlation coefficient between R-R and BP was larger than 0.80. A linear regression was applied to each sequence, and the mean slope was taken as the average of all slopes within supine and upright 5-min periods. This mean slope was considered as depicting the sensitivity of the spontaneous baroreflex (ms.mmHg^{-1}). The effects of stand test and centrifugation on cerebral variables and systemic hemodynamics were assessed by analysis of variance with repeated measures. Student's paired *t* test was used to assess the comparison between pre- versus post-centrifugation for a given posture. Data were expressed as means ± S.D. with levels of $P<0.05$.

3. RESULTS

All subjects experienced the complete $-2\ G_z$ centrifugation without discomfort (nausea, motion sickness…) and each stand test was completed without any presyncopal symptoms. For each stand test, oxy-Hb and total-Hb decreased during standing when deoxy-Hb increased, but these changes were not significantly different from supine (Fig. 1). The decrease in oxy-Hb ($P<0.001$) and in total-Hb ($P<0.05$) upon standing was more important after $-2\ G_z$ centrifugation (Fig. 1). The decrease in upright MAP_{brain} values (Fig. 1) was not different from pre to post $-2\ G_z$. During standing, the change in HR was greater ($P<0.05$) after centrifugation when the decrease in SBS was unchanged (Table 1).

Table 1. Postural changes in systemic hemodynamics during 5-min steady state standing vs. supine[a].

Variable	Pre $-2\ G_z$	Post $-2\ G_z$	P between pre/post $-2\ G_z$
HR	15 ± 10**	19 ± 9**	<0.05
SAP	14 ± 9***	14 ± 14***	NS
DAP	15 ± 6***	17 ± 8***	NS
MAP	15 ± 7***	16 ± 9***	NS
SBS	-7 ± 5*	-10 ± 7*	NS

[a]Values are means ± SD. HR (heart rate) in bpm; SAP (systolic), DAP (diastolic), MAP (mean) arterial pressure in mmHg. SBS indicates spontaneous baroreflex sensitivity in ms.mmHg^{-1}. Significantly different from baseline: *$P<0.01$; **$P<0.001$, ***$P<0.0001$.

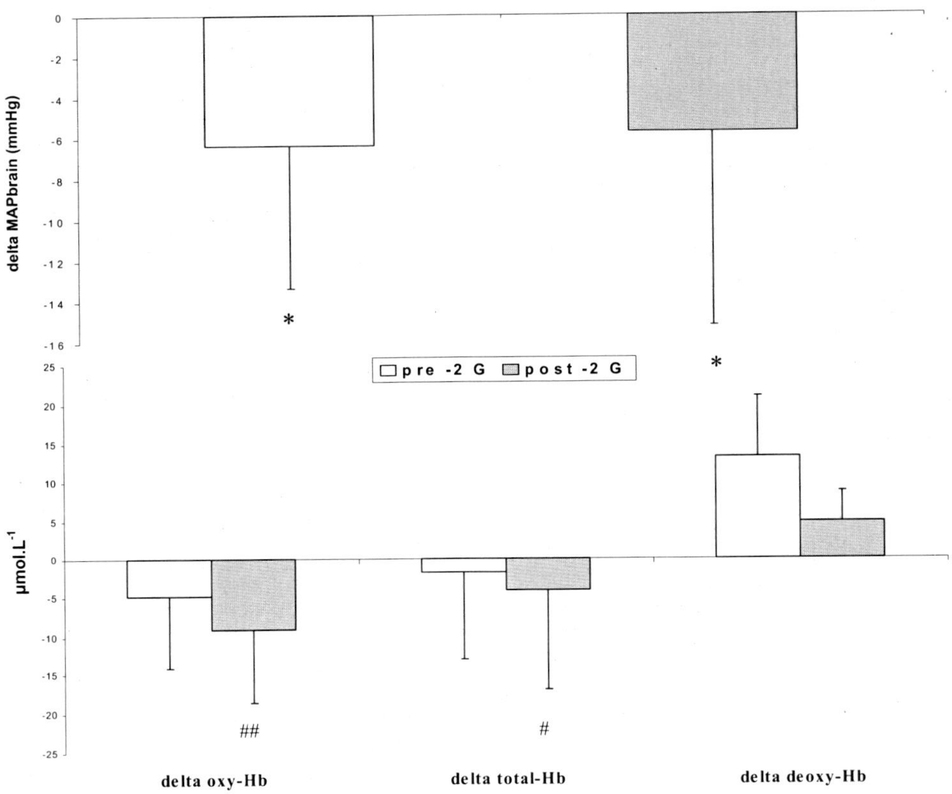

Figure 1. Changes in cerebral oxygenation (oxygenated and deoxygenated hemoglobin, oxy-Hb and deoxy-Hb respectively) and cerebral hemodynamics (mean arterial pressure at brain level and total hemoglobin, MAP_{brain} and total-Hb respectively) during steady standing before (pre) and after (post) a single $-2\ G_z$ centrifugation, with supine control values reference set at zero. *$P<0.05$, significantly different from supine; #$P<0.05$, ##$P<0.001$, significantly different from pre-centrifugation.

4. DISCUSSION AND CONCLUSIONS

The main new finding of the present study was that at steady state, the decrease in brain oxygenation and in cerebral blood volume upon upright posture was larger after $-2\ G_z$ centrifugation. The second interest of our experiment is that it shows that altered cerebrovascular control occurred without decrease in orthostatic tolerance.

Although blood volume measurement cannot be interpreted as cerebral blood flow, it can be assumed that the changes in cerebral oxygenation and cerebral blood volume are indicative of the changes in cerebral perfusion. Indeed, during active standing, cerebral blood flow velocity (CFV) assessed by transcranial Doppler, and cerebral oxygenation assessed by NIRS remained stable for as long as MAP was not affected.[7] And tensing of the leg muscles attenuated the upright reduction in CFV and in cerebral oxygenation.[11] In humans with sympathetic failure,[12] standing induced a more pronounced fall in CFV and in cerebral oxygenation. During head-up-tilt, CFV and cerebral oxygenation both decreased, and presyncopal symptoms related to a further marked drop in oxy-Hb.[13] During LBNP, oxy-Hb and total-Hb started to fall some 20 s prior to the change in HR[14] in subjects with presyncopal symptoms. Furthermore, during LBNP, the decrease in CFV and cerebral oxygenation was greater in spinal cord-injured individuals.[15] We can then assume that in our study, the decrease in cerebral perfusion at steady state upon upright posture was accentuated after –2 G_z centrifugation.

The more pronounced decrease in brain oxygenation and cerebral blood volume upon standing following –2 G_z centrifugation could be explained by a persistent decrease in plasma renin activity in response to the redistribution of the blood from the caudal to the cephalic portion of the body.[16] However in our study, the active standing during the post-centrifugation stand test occurred over 30 min (bioinstrumentation plus 15 min rest and 10 min control in supine position). And it has been demonstrated that plasma renin activity recovered within 15-25 min after 30-min exposure to head-down tilt or water immersion.[16] We can then assume that the decreased plasma renin activity occurring during brief –2 G_z centrifugation does not seem to be a major mechanism in the larger decrease in brain oxygenation and cerebral blood volume that we observed upon standing following centrifugation.

The larger decrease in cerebral perfusion observed in the present study while MAP_{brain} remained within the limits of autoregulation, suggests more likely a rightward or downward shift in the static cerebral autoregulation curve.[1] A rightward shift would have meant that our subjects had continually been operating on the left-hand side of the autoregulation curve, since MAP_{brain} was unchanged after centrifugation. However, as no sign of presyncope has been observed at any time during the stand test, it appears that a downward shift in the curve may be a more likely explanation.[1] A downward shift in the autoregulatory curve has been assumed after parabolic flight in orthostatically intolerant subjects.[17] Indeed, CFV diminished during head-up tilt, despite maintained MAP_{brain}. On the other hand, cerebral hypoperfusion after parabolic flight was correlated to an augmentation of the measured cerebral vascular resistance. One possible mechanism for such an increased cerebral vascular resistance is a local, sympathetically mediated cerebral vasoconstriction. A similar mechanism could be involved after -G_z centrifugation which is known to induce cerebral vasoconstriction to prevent overperfusion.

A greater increase in heart rate was observed with the same decrease in blood pressure during post-centrifugation upright posture in our study. This means that a transient decrease in the sensitivity of the baroreflex has appeared, although not detectable by the spontaneous baroreflex method. Finally, a vestibular activation may have occurred during –2 G_z centrifugation. Indeed during centrifugation, –G_z exposure cannot be pure, since G-transitions effects are unavoidable.[18] And it is well established that vestibular activations induce modulation in the activity of the autonomic nervous system.[19] Taken altogether, we speculate that brief exposure to –2 G_z centrifugation has induced a downward shift in the static autoregulatory curve, which may involve adaptation to cerebral vasoconstriction and/or vestibular activations occurring under –2 G_z exposure.

5. REFERENCES

1. R.L. Bondar, P.T. Dunphy, P. Moradshahi, H. Dai, M.S. Kassam, F. Stein, S. Schneider, and M. Rubin, Vertical shift in cerebral autoregulation curve: a graded head-up tilt study, Can. Aeronaut. Space J. **45**, 3-8 (1999).
2. B.D. Levine, C.A. Giller, L.D. Lane, J.C. Buckey, C.G. Blomqvist, Cerebral versus systemic hemodynamics during graded orthostatic stress in humans, Circulation **90**, 298-306 (1994).
3. R. Zhang, J.H. Zuckerman, and B.D. Levine, Deterioration of cerebral autoregulation during orthostatic stress: insights from the frequency domain, J. Appl. Physiol. **85**, 1113-1122 (1998).
4. J.M. Serrador, S.J. Wood, P.A. Picot, F. Stein, M.S. Kassam, R.L. Bondar, A.H. Rupert, and T.T. Schlegel, Effect of acute exposure to hypergravity (G_x vs G_z) on dynamic cerebral autoregulation, J. Appl. Physiol. **91**, 1986-1994 (2001).
5. C.C. Tran, M. Berthelot, X. Etienne, C. Dussault, J-C. Jouanin, P. Van Beers, A. Serra, and C.Y. Guezennec. Cerebral oxygenation declines despite maintained orthostatic tolerance after brief exposure to gravitational stress. Neurosci. Lett., **380**, 181-186 (2005).
6. M. Ferrari, L. Mottola, and V. Quaresima, Principles, techniques, and limitations of near infrared spectroscopy, *Can. J. Appl. Physiol.* **29**(4), 463-487 (2004).
7. P.L. Madsen, and N.H. Secher, Near-infrared oximetry of the brain, Prog. Neurobiol. **58**, 541-560 (1999).
8. A. Kobayashi, and Y. Miyamoto, In-flight cerebral oxygen status: continuous monitoring by near-infrared spectroscopy, Aviat. Space Environ. Med. **71**, 177-183 (2000).
9. P. van der Zee, M. Cope, S.R. Arridge, M. Essenpries, L.A. Potter, A.D. Edwards, J.S. Wyatt, D.C. McCormick, S.C. Roth, E.O.R. Reynolds, and D.T. Delpy, Experimentally measured optical pathlengths for the adult head, calf and forearm and the head of the newborn infant as a function of interoptodes spacing, Adv. Exp. Med. Biol. **316**, 143-153 (1992).
10. G. Bertinieri, M. Di Rienzo, A. Cavallazzi, A.U. Ferrari, A. Pedotti, and G. Mancia, Evaluation of baroreceptor reflex by blood pressure monitoring in unanesthetized cats, Am. J. Physiol. **254**, 377-383 (1988).
11. J.J. van Lieshout, F. Pott, P.L. Madsen, J. van Goudoever, and N.H. Secher, Muscle tensing during standing, Effects on cerebral tissue oxygenation and cerebral artery blood velocity, Stroke **32**,1546-1551 (2001).
12. M.P.M. Harms, W.N.J.M. Colier, W. Wieling, J.W.M. Lenders, N.H. Secher, and J.J. van Lieshout, Orthostatic tolerance, cerebral oxygenation, and blood velocity in humans with sympathetic failure, Stroke **31**, 1608-1614 (2000)
13. K. Krakov, S. Ries, M. Daffertschofer, and M. Hennerici, Simultaneous assessment of brain tissue oxygenation and cerebral perfusion during orthostatic stress, European Neurology **43**, 39-46 (2000).
14. D.H. Glaister, and N.L. Miller, Cerebral tissue oxygen status and psychomotor performance during lower body negative pressure (LBNP), Aviat. Space Environ. Med. **61**, 99-105 (1990).
15. S. Houtman, J.M. Serrador, W.N.J.M. Colier, D.W. Strijbos, K. Shoemaker, and M.T.E. Hopman, Changes in cerebral oxygenation and blood flow during LBNP in spinal cored-injured individuals, J. Appl. Physiol. **91**, 2199-2204 (2001).
16. M. Shiraishi, M. Schou, M. Gybel, NJ. Christensen, and Peter Norsk, Comparison of acute cardiovascular responses to water immersion and head-down tilt in humans, J. Appl. Physiol. **92**, 264-268 (2002).
17. J.M. Serrador, J.K. Shoemaker, T.E. Brown, M.S. Kassam, R.L. Bondar, and T.T. Schlegel, Cerebral vasoconstriction precedes orthostatic intolerance after parabolic flight, Brain Res. Bull. **53**, 113-120 (2000).
18. B. Cheung, and WA. Bateman, G-transition effects and their implications, Aviat. Space Environ. Med. **72**, 758-762 (2001).
19. B.J. Yates, Vestibular influences on the sympathetic nervous system, Brain Res. Rev. **17**, 51-59 (1992).

NORMAL CARDIAC OUTPUT, OXYGEN DELIVERY AND OXYGEN EXTRACTION

Christopher B Wolff *

Abstract: The total amount of blood flow circulating through the heart, lungs and all the tissues of the body represents the cardiac output. Most individual tissues determine their own flow in proportion to their metabolic rate. The skin is a notable exception where the priority is thermal rather than metabolic. Renal blood flow and metabolic rate are related but plasma flow determines metabolic rate rather than metabolic rate determining blood flow.[1] Brain, heart, skeletal muscle and the splanchnic area all vary their blood flows according to local tissue metabolic rate. Summation of peripheral blood flows constitutes venous return and hence cardiac output. Cardiac output is therefore, largely, determined by the metabolic rate of the peripheral tissues; the heart '*from a flow standpoint, plays a "permissive" role and does not regulate its own output*'.[2] This peripheral tissue, largely metabolic, determination of cardiac output has been known for many years.[3,4]

Evidence will be presented that blood flow is scaled according to a tissue specific ratio of oxygen delivery (DO_2) to oxygen consumption (VO_2). For the brain DO_2 is approximately three times VO_2, for heart muscle DO_2 is 1.5 to 1.6 times VO_2 and is very similar for skeletal muscle for moderate exercise. Brain, heart and skeletal muscle have the ability to sustain appropriate blood flow in the face of varying blood pressure within limits – the phenomenon known as 'autoregulation'. "Autoregulation, in regard to arterial blood pressure, has been observed" also "in the kidney" and "modest autoregulation" was observed "in the intestines and liver but not in skin".[3] Guyton et al. [4] have suggested that the term 'auto-regulation' should also include variation of blood flow in proportion to metabolic rate and the compensatory changes in blood flow which occur in the face of varying arterial oxygen content (CaO_2). This article gives examples of the very precise compensation for CaO_2 change in the form of sustained tissue specific $DO_2{:}VO_2$ ratios (corresponding with tissue specific oxygen extraction, $E = VO_2/DO_2$). The adequacy of

* Applied Physiology, Block 9, St Thomas's Hospital, Lambeth Palace Rd., London, SE1 7EH, UK and Clinical Pharmacology, William Harvey Research Institute, Barts, and The London, Charterhouse Square, London, EC1M 6BQ, UK

this adjustment for brain, exercising skeletal muscle and heart is particularly striking; skeletal muscle will, for example when CaO_2 is reduced, steal blood supply from non-exercising tissues sustaining its own oxygen delivery at normal levels.[5]

1. INTRODUCTION

For the commonest man-made circulatory systems the energy input to the pump determines the pump output. In contrast the circulation of blood in man and other animals is regulated peripherally, from adjustments made in a circulation where "both the pipe system and the fluid are able to change characteristics from time to time and thereby impose varying demands on the pump. Since the main changes in cardiac output are determined by the metabolic requirements of the body and not by voluntarily switching the pump to a higher setting, it is evident that the body sets the pace and not the pump. Thus, the heart, from a flow standpoint plays a permissive role and does not regulate its own output. So in the human circulation, the heart functions as a 'demand' pump ......".[2]

Guyton et al.[4] in their 1973 monograph quoted work dating from the early 1900s which demonstrated that:

1. "Cardiac output almost exactly parallels the level of metabolism....", and
2. "On the basis of these studies, it has now become widely recognised that the single factor most responsible for the significant linkage between metabolic rate and cardiac output is the tissue need for oxygen."

The verified specificity for oxygen is based on extensive experimentation. Guyton et al.[4] quote 21 references dating from 1922 to 1971 and at the end of the list of references comments: "... and many others."

Measurements of cardiac output prior to the use of dye dilution and the Fick method, on average, gave answers which were about 2/3 of the correct values.[4]

Below are listed the main, established features of circulatory control (other than blood pressure regulation) including metabolic and other contributions to venous return and hence cardiac output. The permissive range, mentioned in point A, is the range over which the heart will act in its, so-called, "permissive" role, simply passing on the peripherally determined venous return as cardiac output. The extra capacity for pumping above the resting value has been referred to as the "cardiac reserve" and may be 25 l min^{-1} or more for a highly trained athlete.[6]

A. The heart (in the permissive range) puts out what it receives

B. Peripheral tissues demand blood flow appropriate to:

1. Metabolic rate (VO_2): – Whole body exercise, skeletal muscle, heart, splanchnic (including the liver) and brain
2. Renal priorities (electrolyte regulation)
3. Skin (thermal priorities)
4. Bone (priorities unknown – magnitude of blood flow hard to judge)
5. Artificial shunt

C. Autoregulation – appropriate tissue blood flow

1. Flow sustained despite blood pressure change (within limits) confirming peripheral regulation.
2. The sum of all peripheral blood flows = venous return = cardiac output (within permissible range)

D. Oxygen's importance as the specific index of metabolic rate, e.g.

1. Reperfusion blood flows repay the oxygen debt for brief occlusions.[4]
2. Cardiac output is linearly related to oxygen consumption.

The features listed will be illustrated by examples from the literature.

2. EVIDENCE FROM THE LITERATURE

2.1. The Heart Puts Out What It Receives

Donald and Shepherd [7] studied exercise in two groups of dogs, one entirely normal, and the other after recovery from cardiac denervation. The results are illustrated in Figure 1.

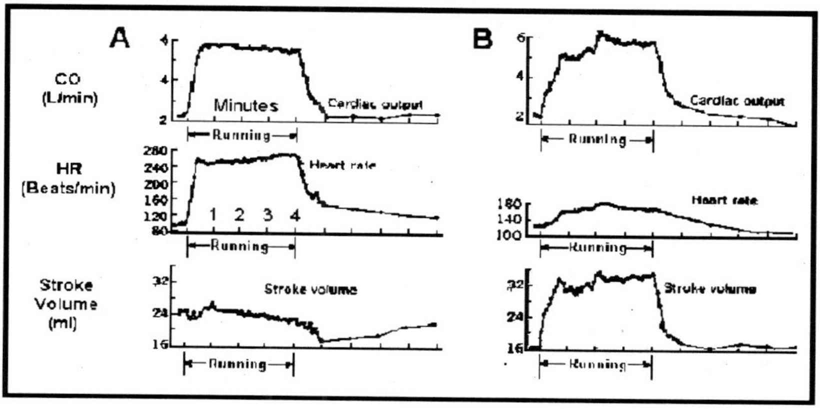

Figure 1. Results of exercise (running) in normal dogs are shown in A, where the cardiac output increase is mainly due to increased heart rate. This contrasts with the cardiac denervated dogs (B) where most of the increased cardiac output resulted from increased stroke volume. Used with permission from D. E. Donald and J. T. Shepherd, Initial cardiovascular adjustment to exercise in dogs with chronic cardiac denervation, *Am J. Physiol.* 207(6), 1325-1329 (1964).

The cardiac denervated dogs showed a cardiac output much the same as occurred in the intact dogs, largely achieved via an augmented stroke volume. This illustrates feature A above "the heart puts out what it receives".

A second example illustrating that the heart will pump out what it receives is the situation where an artificial shunt is created. Here there is no metabolic rate priority or other control. Blood flow depends simply on the resistance to flow and arterial blood pressure; cardiac output keeps up precisely with the excess inflow unless it is so large that the cardiac reserve is insufficient.[4]

2.2. Blood Flows Related To Oxygen Consumption

2.2.1. Whole Body and Skeletal Muscle in Moderate Exercise

The linear relationship between cardiac output and oxygen consumption has been known for many years [4,6,8,9] and is illustrated in Figure 2 (left hand panel). On the right the schematic allows the deduction that if the excess cardiac output represents the blood flow to muscle then the muscle arterio-venous difference is 1/B, where B is the slope of the line.[5]

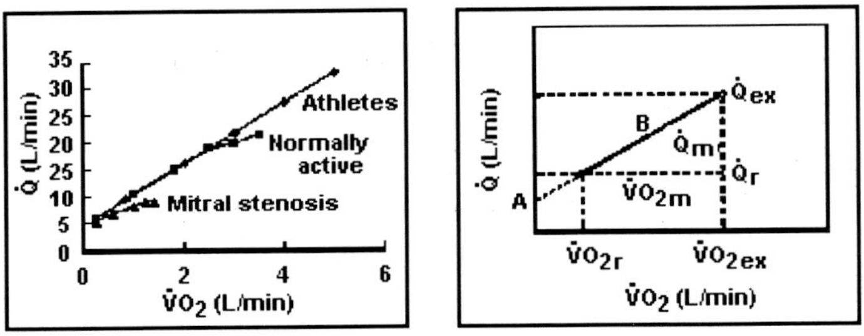

Figure 2. On the left is a classic illustration of the way cardiac output increases with oxygen consumption (after Rowell [8,9]) and on the right a hypothetical arrangement whereby the increased cardiac output above the resting value represents blood flow to the exercising muscle.[5] Key: Q, cardiac output; A, intercept; B, slope; VO_{2r} and VO_{2ex}, resting and exercising VO_2; Q_r and Q_{ex}, resting and exercising cardiac output; Q_m, muscle blood flow; VO_{2m}, muscle oxygen consumption.

Using the hypothesis illustrated in Figure 2 on the right most standard figures given for the relationship between cardiac output and oxygen consumption, dating from the early methods, gave oxygen extraction values which were too high (90% to 100% or more!). The cardiac output measurements were too low (respiratory methods). Analysis of data from Koskolou et al.,[10] where cardiac output was accurate (measured by dye dilution), allowed calculation of muscle blood flows from the Fick equation because arterial and muscle venous oxygen content were available. When the subjects had a normal arterial oxygen content (CaO_2) muscle blood flow was equal to the excess cardiac output above the resting value. Furthermore, oxygen extraction, E, was 68% (close to 2/3) for 30 watts (w) and 73w exercise. When the subjects were made normovolaemically anaemic skeletal muscle oxygen extraction was still close to 2/3; muscle blood flow had increased enough to sustain precisely the DO_2 required. However, the total cardiac output was now insufficient to supply the original resting blood flow to the non-exercising tissues – there was therefore now 'steal' of blood flow by the exercising muscle thereby supplying the required excess muscle blood flow.[5]

Figure 3 illustrates the findings; on the left exercising muscle blood flow was greater in the anaemic subjects for any given metabolic rate than in the normal subjects; this resulted in the anaemic subjects having the same oxygen delivery for any given wattage as normal subjects (middle panel). The right hand panel shows how oxygen delivery and oxygen consumption vary with the work rate, illustrating the high oxygen extraction (similar to cardiac muscle).

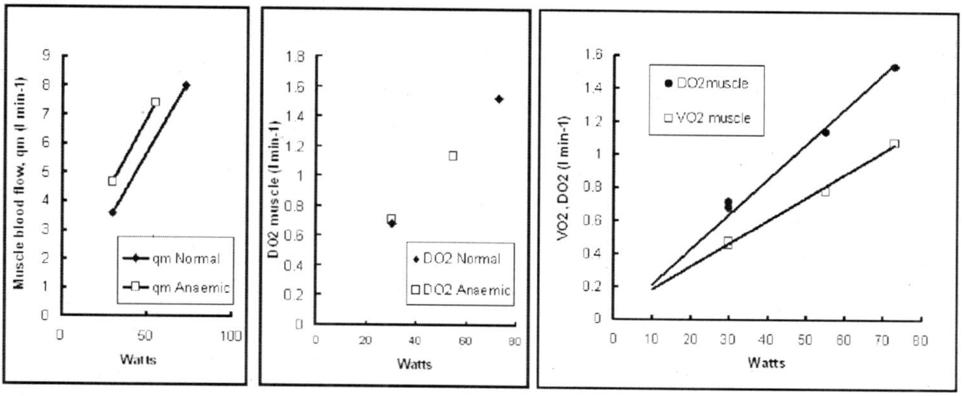

Figure 3. Exercising muscle blood flow (left panel) is greater in anaemia than normal, compensating for the reduced CaO_2 (middle panel) so that muscle oxygen delivery is the same for any given metabolic rate as it is with normal CaO_2. The right hand panel shows that DO_2 is about 1.5 times VO_2; expressed another way, oxygen extraction, E, is 2/3 of oxygen delivery. Analysis[4] of Koskolou et al. data.[10]

These results embody an important feature of the circulation. For this specific tissue a constant oxygen DO_2:VO_2 ratio and oxygen extraction have been maintained over a moderate range of metabolic rate both with normal and with reduced CaO_2 (in this case from anaemia). The same skeletal muscle oxygen extraction (around 2/3) was also present at 30w where CaO_2 was lowered with hypoxia, anaemia and combined hypoxia and anaemia in subjects studied by the same group.[11] Constant oxygen extraction over this metabolic rate range also means that there will be a specific rate of oxygen delivery at a particular metabolic rate.

Exercising skeletal muscle thus determines its own appropriate blood flow, so long as it is in a moderate exercise range (in heavy exercise oxygen extraction is increased) and so long as CaO_2 is above a very low value (in this case DO_2 was sustained at the lowest recorded CaO_2 – 11.47 ml per 100 ml blood, 60% of the normal value).[5,10,11]

The fact that blood flow to other organs was reduced and hence oxygen delivery to other organs (probably splanchnic) was compromised in the anaemic subjects of Koskolou et al. [10] shows that the autoregulation for exercising skeletal muscle is more powerful than for the other organs.

2.2.2. Cerebral Blood Flow and Oxygen Delivery in Hypoxia

Cerebral auto-regulation of oxygen delivery in the face of hypoxia is illustrated by an analysis of a study by Severinghaus et al.[12] The analysis examined the cerebral oxygen delivery from the Severinghaus et al. study, firstly over the first 8 hours,[13] and in a second study extending the time base to include findings at 3 days and 5 days.[14] The fall in

$PaCO_2$ was slight at 8 hrs with a maximal fall in SaO_2 at that time (Figure 4, left hand panel) and blood flow was maximal (an approximately 25% increase, middle panel).

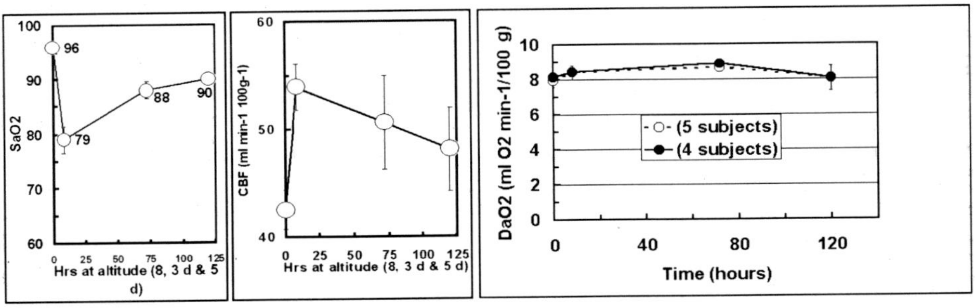

Figure 4. The left hand and middle panels show arterial saturation and cerebral blood flow (CBF) at sea level and successively after arrival at 3810 m altitude. Values are plotted for 8 hours, 3 days and five days. The large initial fall in SaO_2 is mirrored by a 25% rise in CBF. The resultant oxygen delivery (DaO_2) values are shown in the larger right hand panel. Mean values are shown for 4 subjects and for 5 subjects (values either homogeneous or not). Cerebral DaO_2 was well sustained.[13,14] Analysis of data from Severinghaus et al.[12]

With acclimatization, $PaCO_2$ fell further, SaO_2 improved and cerebral blood flow declined gradually over the next five days (Figure 4 middle panel). Oxygen delivery (right hand panel of Figure 4) was much the same as at sea level throughout the five days at altitude.[14] Hence; normal oxygen delivery was sustained throughout.

The proportion of oxygen extracted from that delivered to the brain as a whole (global oxygen extraction, E) is considerably less than for heart and skeletal muscle, about 34%, equivalent to a DO_2:VO_2 ratio of approximately 3:1.

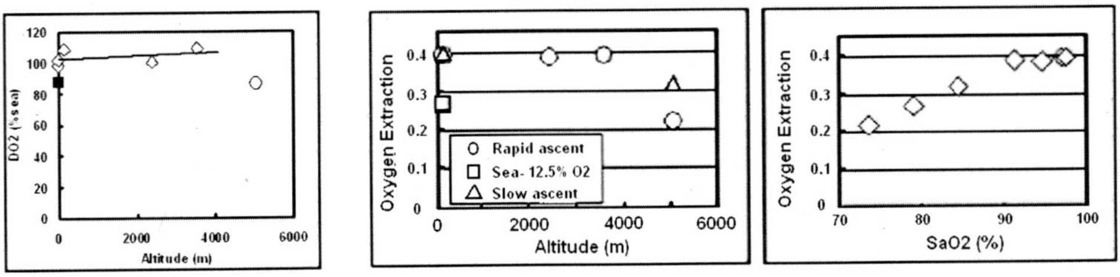

Figure 5. Cerebral oxygen delivery, as a percentage of the sea level value is shown for 2400 m, 3550 m and 5050 m (left hand panel). DO_2 is decreased at 5050 m (semi-acute ascent) but is the same as at sea level for the lower two altitudes.[18] ■ ..... Inspiration of 12.5% oxygen at sea level. The middle and right hand panel show estimates of cerebral oxygen extraction (E) for a region of frontal cortex, derived from combining near infra-red spectroscopy (NIRS) values of rSO_2 (ratio of oxygenated to total Hb) with SaO_2. The middle panel shows E against altitude and closely resembles the pattern for cerebral DO_2 with an extra point from a slow ascent to high altitude (Δ, 5000 m). The same values of E are shown against SaO_2 in the right hand panel.[19]

Measurements of middle cerebral artery velocity and SaO_2 made by Imray et al. during ascent to several different altitudes [15,16,17] allow calculation of values proportional to DO_2. Normal DO_2 was sustained for moderately high altitudes (2400 m and 3550 m) but was reduced to around 85% of the sea level value at 5050 metres (Figure 5, left hand panel).[18] DO_2 was reduced for acute hypoxia at sea level (black square in the figure). Cerebral DO_2 was the same as at sea level when SaO_2 values were above 90%.

Near infra-red spectroscopy (NIRS) measurements over the frontal area were also made by Imray et al. [15,16,17] The ratio of oxygenated to total blood in the sample area, rSO_2, lumps arterial and venous blood together; but analysis of the results in combination with SaO_2 suggest that oxygen extraction and the ratio of arterial to venous blood volume were unchanged over the same range as that for which middle cerebral artery DO_2 was constant.[18] This allowed derivation of oxygen extraction, E, for the region.[19] Cerebral cortical E, derived from NIRS measurement, is presented for the data from Imray et al. [15,16,17] in the middle and right hand panels of Figure 5. E plotted against altitude (middle panel) is very similar to oxygen delivery against altitude (left hand panel). The extra point (triangle, middle panel) represents the mean value for E from 14 subjects who had ascended slowly to 5000 m (taking around 17 days); E is less reduced than the value for the semi-acute ascent to much the same altitude. Adjustments are completely compensatory (constant E) for SaO_2 values above 90% (Figure 5 right hand panel) but appear to become progressively more abnormal for lower SaO_2.

In summary, cerebral oxygen delivery and extraction are sustained at the normal (sea level) value in the face of hypoxic reductions in SaO_2 above (approximately) 90%. Cerebral blood flow and DO_2 are therefore regulated by intrinsic cerebral mechanisms.

2.2.3. Myocardial Oxygen Extraction

Constancy of myocardial oxygen extraction in the face of increased cardiac output means that there is normally an increase in coronary blood flow in proportion to myocardial VO_2. Where there is also low CaO_2 (hypoxia or anaemia) the situation is more complicated. There is, nevertheless, still a priority to sustain myocardial DO_2 appropriate to the metabolic rate.

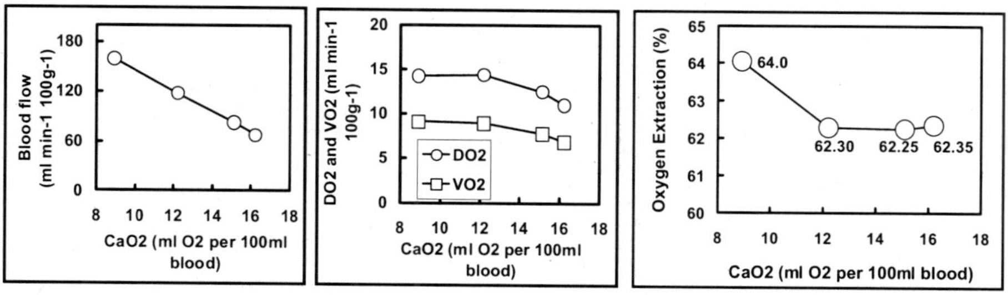

Figure 6. Myocardial (right ventricular) blood flow (left hand panel), oxygen delivery and consumption (middle panel) and oxygen extraction (right hand panel) from Martinez et al.[20] show the effect of hypoxia in dogs. Hypoxia changes myocardial blood flow partly as a result of the hypoxia and partly increased myocardial oxygen consumption due to increased cardiac output. Myocardial oxygen extraction (right hand panel) is constant for CaO_2 above 12 ml. O_2 $(100\ ml)^{-1}$ blood; equivalent to a ratio of DO_2 to VO_2 of 1.6 (approx.).

Constancy of myocardial oxygen extraction in the face of changes in both metabolic rate and arterial oxygen content can be illustrated from the results of Martinez et al.[20]. The experiment consisted of hypoxic challenges in instrumented conscious dogs. Figure 6 shows blood flow values for the normal state and three hypoxic states (left hand panel).

Myocardial oxygen consumption, shown in the middle panel, is increased in hypoxia because increased cardiac output sub-serves the needs of the rest of the body (compensatory increases in blood flows elsewhere). Similarly, myocardial oxygen delivery is increased, also shown in the middle panel. The right hand panel shows that oxygen extraction is barely changed from normal for two of the hypoxic states. Only with an arterial oxygen content of 9 ml. O_2 $(100\ ml)^{-1}$ blood was there an increase in oxygen extraction.

The heart, like exercising skeletal muscle and brain, determines its own blood flow and specific $DO_2:VO_2$ ratio (here 1.6, oxygen extraction 62.3%) in the face of changes in myocardial oxygen consumption and arterial oxygen content. There is a lower limit for this adjustment at a CaO_2 value between 8 and 12 ml O_2 per 100 ml blood.

2.2.4. Splanchnic Oxygen Extraction

The splanchnic area (gastro-intestinal tract, pancreas, spleen and liver) drains firstly into the hepatic portal vein and then the liver (see Figure 7). In the figure 'gut' refers to gastro-intestinal tract, pancreas and spleen. The oxygen extraction and $DO_2:VO_2$ estimates have been made from experimental details given in the text by Folkow and Neil.[21]

Oxygen extraction is only about 16% for the portal system and 21% overall (including the liver). These correspond with $DO_2:VO_2$ ratios of around 20%. Blood flows in the figure are for the time between meals. After food flows increase in line with the increased metabolic rate, due to secretion, absorption and smooth muscle contraction. There are specific increases in local blood flows at the sites of increased oxygen consumption.

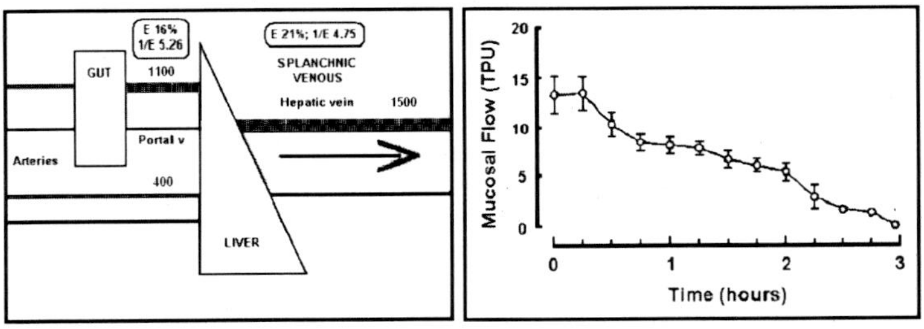

Figure 7. Left hand panel: schematic diagram of the splanchnic blood flow and arterial and venous oxygen content values (clear, oxygenated blood; dark stippled, de-oxygenated blood), with blood flow and oxygen extraction estimates (derived from Folkow and Neil [21]). Portal oxygen extraction is less than overall splanchnic oxygen extraction. Normal splanchnic function is associated with low oxygen extraction (a high oxygen delivery to consumption ratio – around 5). The portal blood flow (1100 ml min^{-1}) and hepatic vein blood flow (1500 ml min^{-1}) will be greater after meals. Right hand panel: Ileal mucosal blood flow in dogs subject to controlled haemorrhage – 5 ml kg^{-1} every 15 minutes. The fall in blood flow after the first increment is very early and occurs with very little overall haemo-dynamic change. Intestinal mucosal autoregulation is far less robust than for brain, heart and skeletal muscle.[22]

The "modest auto-regulation" observed "in the intestines and liver" to quote Green et al.[3] can be illustrated by the right hand panel of Figure 7. The modest increments of controlled haemorrhage (5 ml kg^{-1} every 15 minutes) caused a fall in mucosal blood flow very early, after only the second increment.[22]

The splanchnic area, with high $DO_2:VO_2$ ratios is another example of blood flows related to oxygen consumption but appears to be vulnerable to requirements for blood flow elsewhere. Details of the response of blood flow to reductions in CaO_2 are not readily available but it is anticipated that the 'steal' observed with modest exercise in subjects with reduced CaO_2 will mean that the tolerable range of CaO_2 reduction will be less than for skeletal muscle or brain.

2.3. Oxygen Consumption Related to Blood Flow – The Kidney

Experimental reductions in renal blood flow cause a proportional reduction in renal oxygen consumption and hence no change in oxygen extraction.[1] The metabolic rate depends on sodium absorption which depends on blood flow, in particular, plasma flow. Quoted values for renal oxygen extraction are low; approximately 1/6 appears to be a consensus (16-17%) so the DO_2:VO_2 ratio is normally about 6:1. When the oxygen content of the arterial blood is reduced there is little change in renal plasma flow so the renal oxygen consumption remains much the same. Oxygen delivery is now lower for any given plasma flow than with normal CaO_2. Hence, oxygen extraction is increased and erythropoietin is produced.

There is still a correlation between blood flow and metabolic rate for the kidney in normal subjects (normal CaO_2), so the correlation for the whole body still holds good. For the kidney however the causal relationship (flow determines VO_2) is the reverse of that found in the other major organs – skeletal muscle, heart, brain and the splanchnic area.

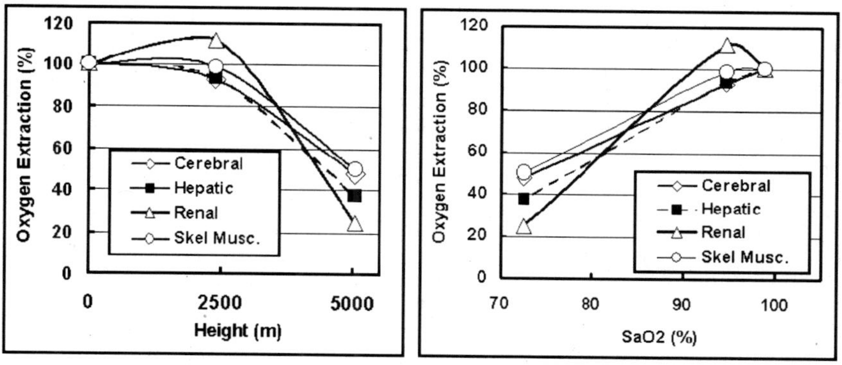

Figure 8. Oxygen extraction values (E) are shown for brain, liver, skeletal muscle and kidney in resting subjects at sea level, 2400 meters and 5050 meters altitude after moderately rapid ascent. E was derived from NIRS measurement (rSO_2, oxygenated/total Hb) and SaO_2: $E = 1.39x(1 - rSO_2/SaO_2)$. Only the renal value of E is increased at 2400 meters – left hand panel. The right hand panel shows the same values for oxygen extraction plotted against arterial oxygen saturation. All values were reduced at 5050 meters. After Wolff et al.[19]

The effect of hypoxia on renal oxygen extraction, E, is illustrated in Figure 8, where E has been calculated from rSO_2 and SaO_2 for brain, liver, skeletal muscle and kidney at rest. The NIRS measurements made by Imray et al.[15,16,17] included rSO_2 over these organs with calculation according to $E = 1.39x(1 - rSO_2/SaO_2)$ [19] to give values at sea level, 2400 metres and 5050 metres; measurements were made after a moderately rapid ascent. Figure 8 shows that E for brain, liver and resting skeletal muscle was unchanged at 2400 meters but renal oxygen extraction was increased. All E values were reduced at 5050 meters.

Renal perfusion depends, not on metabolic rate, but on electrolyte priorities. Oxygen extraction however is normally sustained at a low value and “auto-regulation in regard to arterial blood pressure has been observed in the kidney".[3] Although perfusion is not dependent on the metabolic rate the constancy of blood flow in the face of blood pressure changes suggests peripheral control, in this case dependent on electrolyte regulation.

3. DISCUSSION

3.1. Tolerable range for constant oxygen extraction is CaO_2 dependent

From the exercising muscle study (section 3.2.1) it appears that the sustained muscle $DO_2:VO_2$ ratio (and oxygen extraction) is specifically related to CaO_2 rather than to PO_2 or Hb. The values of oxygen tension in the exercising muscle venous blood were all very similar for normal and anaemic states but those for 30-watt exercise in hypoxia were reduced as seen in Figure 9.

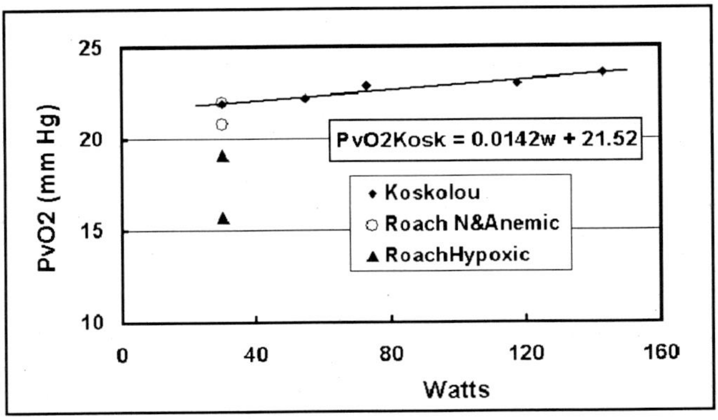

Figure 9. Exercising muscle venous PO_2 (P_VO_2) for the normal and anaemic subjects of Koskolou et al.[10] were all on a shallow line whereas the hypoxic values from Roach et al.[11] were well below the anaemic ones.

The dependence of constant oxygen extraction (and $DO_2:VO_2$ ratio) on CaO_2 rather than PO_2 or Hb suggests the importance of maintaining an oxygen supply. However, the mechanism driving the oxygen into the mitochondria depends upon PO_2 and the diffusion rate will depend on distance as well as the diffusion coefficient for oxygen. Krogh's work showed very large changes in inter-capillary distances across a range of skeletal muscle metabolic rates which allow adequate diffusion rates.[23] The diffusion coefficient for oxygen is probably constant, though some augmentation of the diffusion coefficient above that for water has been shown to occur with artificial dextran solutions.[24] A feature of the diffusion of oxygen suggested by this is the possibility that there could be augmentation of the diffusion coefficient from alterations in tissue polymer changes. Also important is the suggestion that diffusion of oxygen starts at arteriolar level as this may allow local tissue signalling.[25]

3.2. Possible reduction of resting tissue VO_2 when CaO_2 is excessively low

The finding that there is both reduced oxygen delivery and extraction in brain at SaO_2 values below 90% (section 3.2.2) is either an artefact of the calculation or means that there is reduced oxygen consumption. This should be testable by modern methods and would be of considerable interest. This appears to apply also to liver, kidney and resting skeletal muscle (section 3.3, Figure 8). It is possible that lowering of VO_2 occurs when tissue self-regulation is beyond the range where adequate oxygen delivery can be sustained in tissues not being driven to higher metabolic rates. Very high oxygen extraction values are known to occur in exercising skeletal muscle where the metabolic rate is driven to high and maximal levels. Here the metabolic rate is driven to the high levels in contrast to resting tissues.

3.3. Effects on mechanisms sustaining blood flow adjustments

The impaired intestinal perfusion in haemorrhage in dogs (illustrated in Figure 7; section 3.2.4) was improved significantly by the administration of a dopamine-1 agonist (Figure 10, left hand panel). Guzman et al.[22] found no difference in VO_2 suggesting an improved DO_2:VO_2 ratio. This is compatible with E,control at baseline of 0.24 and E,drug of 0.21 (DO_2:VO_2 ratios: 4.2 and 4.8, respectively).

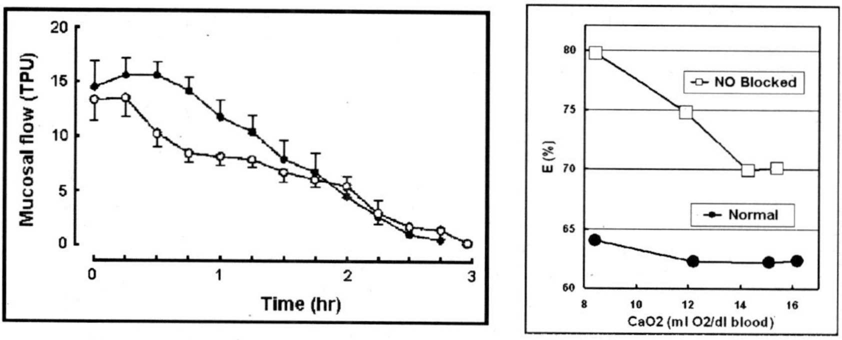

Figure 10. The left hand panel here shows improved intestinal mucosal blood flow in dogs, haemorrhaged 5 ml Kg^{-1} every 15 minutes, on a Dopamine-1 (DA-1) receptor agonist (•). With no drug blood flow is significantly worse for the first 6 bleeding episodes. TPU .... Tissue perfusion units. There was no change in mucosal VO_2.[22] In the right hand panel the normal, well-sustained cardiac oxygen extraction (Figure 6) is grossly modified by NO blockade. Oxygen extraction is greater and increases over most of the hypoxic range.

It can be seen in Figure 10 (right hand panel) that the well-sustained myocardial oxygen extraction in hypoxia (Figure 6 right hand panel and lower section above) is radically altered by nitric oxide (NO) blockade. Oxygen extraction is higher and not well-sustained. If NO is basic to the control of blood flow for other metabolic dependent tissues this will be a combination of effects on peripheral tissues as well as the heart.

3.4. Arterial blood pressure control

To quote Guyton et al.[4] "If each local tissue is to control its own blood flow by dilating or constricting its blood vessels, then it is essential that the arterial pressure be maintained at a pressure high enough to supply the blood demanded by the tissues." And, " it is often thought that nervous control of arterial pressure is effected almost entirely by changing the resistance to blood flow through the small tissue blood vessels. However, an increase in resistance would decrease blood flow through the tissues. Therefore, when a tissue demands additional quantities of blood flow because of its local needs, vasoconstriction of its arterioles to maintain the arterial pressure would be very detrimental to the tissue's own needs. Fortunately, the local vasodilating effect caused by increased metabolism usually prevails over the vaso-constricting effect of the nervous signals, and the net result is decreased vascular resistance rather than increased resistance. Yet, how can the nervous system maintain arterial pressure in the face of the decreasing vascular resistance? The answer is: by increasing the filling pressure of the circulation. Strong sympathetic stimulation of the large vessels, particularly the veins, can increase this pressure, called the *mean systemic pressure,*[1] from a normal value of 7 mm Hg up to 18 to 21 mm Hg, almost a three fold increase. This is essentially the same effect as increasing the pressures in all the peripheral vascular reservoirs by an average of three fold, which greatly increases venous return and cardiac output. Therefore, the arterial pressure is maintained by an increase in cardiac output and not by an increase in peripheral resistance." As an example he quotes experimental work where heavy exercise in the sympathectomized-vagotomised dog led to a gross fall in blood pressure, whereas in the intact animal there was an increase in blood pressure.

3.5. Overview

Determination of blood flows to specific organs and the overall (whole body) proportional relationship of cardiac output to metabolic rate confirms that it is tissue requirements, not cardiac function per se which determine cardiac output.

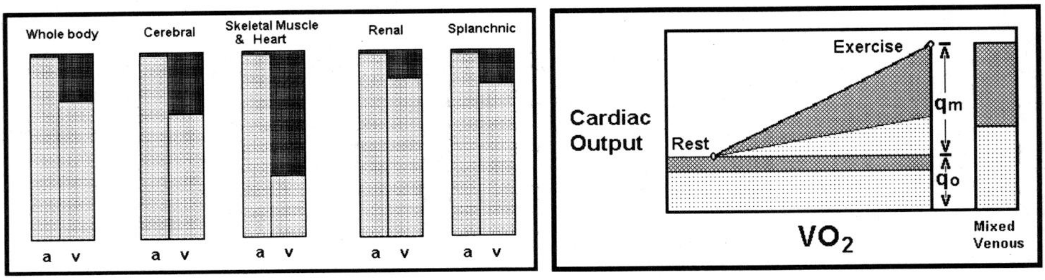

Figure 11. The panel on the left shows normal arterial and venous oxygenation for the body as a whole, for the brain, skeletal muscle and heart, the kidney and the splanchnic area. The venous oxygenation for the whole body is as shown at rest, approximately ¼ oxygen extraction, cerebral 1/3, skeletal muscle and heart approximately 2/3, kidney about 1/6 and the splanchnic area around 1/5. In moderate exercise skeletal muscle removes 2/3 of arterial oxygen but the rest of the body still removes ¼ as illustrated in the panel on the right. This depicts venous oxygenation for muscle (blood flow qm) and the rest of the body (blood flow qo). These will mix to give a combined, mixed venous blood with progressively lower oxygen saturation as exercise intensity increases.

Figure 11 illustrates the oxygen extraction status for the various major organ systems and also for the whole body at rest in the left hand panel. The venous value for the whole body at rest, representing the total of all tissue blood flows is approximately ¼ desaturated. In exercise the exercising muscle venous drainage, with 2/3 desaturation, is

[1] 'mean systemic pressure' is the effective mean pressure which would obtain if blood flow was stopped and the blood volume distribution had reached equilibrium.

added in progressively higher proportion (with heavy exercise desaturation is even greater). The right hand panel illustrates admixture of the 2/3 desaturated muscle venous blood to the ¼-desaturated blood from the rest of the body. The mixed venous saturation, $SvO_2 = (qm \times 1/3 + qo \times 3/4) / \text{Cardiac output}$, for the modest exercise range.

In summary, blood flows are largely determined at tissue level and normally sustain a tissue specific oxygen extraction. This tissue specific oxygen extraction is sustainable within limits of arterial blood pressure (autoregulation), metabolic rate and arterial oxygen content. The CaO_2 range over which a normal oxygen extraction can be sustained varies from one tissue group to another; skeletal muscle and heart sustain normal extraction (approximately 2/3, DO_2:VO_2 of 1.5 to 1.6) down to an oxygen content at least as low as 60% of normal, cerebral oxygen extraction 1/3 (DO_2:VO2 of 3) above 90% or so for hypoxia, the limit for anaemia may well be at the same CaO_2. For the kidney and splanchnic areas oxygen extraction is much lower (1/6 to 1/5, DO_2:VO_2 ratio of 5 or 6) and the splanchnic area does not appear to be able to stand low CaO_2 or blood pressure. Limits for the kidney are not yet clear but it extracts more oxygen when CaO_2 is low rather than sustaining a constant value.

4. REFERENCES

1. H. Valtin, *Renal Function: Mechanisms Preserving Fluid and Solute Balance in Health and Disease.* Ch 6 Renal Hemodynamics and Oxygen Consumption. (Little, Brown and Company, Boston, 1973), pp 177-196.
2. A. J. Carlson and V. Johnson, *The Machinery of the body*, Third edition. (The University of Chicago Press, Chicago, 1948).
3. H. D. Green, C. E. Rapela and M. C. Conrad, Resistance (conductance) and capacitance phenomena in terminal vascular beds. In, *Handbook of Physiology*, *Circulation,* (Am. Physiol. Soc., Washington D.C, sect. 2, vol. II, chapter. 28, 1963), pp. 935-960.
4. A. C. Guyton, C. E. Jones and T. G. Coleman, *Circulatory Physiology: Cardiac Output and its Regulation,* (W. B. Saunders Company, Philadelphia, 1973).
5. C. B. Wolff. Cardiac output, oxygen consumption and muscle oxygen delivery in submaximal exercise: normal and low O_2 states, *Adv. Exp. Med. Biol.* **510**: 279-284 (2003).
6. A. C. Guyton, Integrated dynamics of the circulation and body fluids. Ch 6, In, *Pathologic Physiology: Mechanisms of Disease*, edited by W. A. Sodeman and T. M. Sodeman, (W. B. Saunders Co., Philadelphia and London, 1979), pp 169-197
7. D. E. Donald and J. T. Shepherd, Initial cardiovascular adjustment to exercise in dogs with chronic cardiac denervation, *Am J. Physiol.* **207**(6), 1325-1329 (1964).
8. L. B. Rowell, *Human Circulation: Regulation During Physical Stress* (OUP *Oxford*,. 1986).
9. L. B. Rowell, *Human Cardiovascular Control* (OUP Oxford., 1993).
10. M. D. Koskolou, R. C. Roach, J. A. Calbet, G. Rådegran, and B. Saltin, Cardiovascular responses to dynamic exercise with acute anemia in humans, *Am. J. Physiol.* **273**, H1787-H1793 (1997).
11. R. C. Roach, M. D. Koskolou, J. A. L. Calbet, and B. Saltin, Arterial O_2 content and tension in regulation of cardiac output and leg blood flow during exercise in humans, *Am. J. Physiol.* **276**, H438-H445 (1999).
12. J. W. Severinghaus, H. Chiodi, E. I. Eger, B. Brandstater, and T. F. Hornbein, Cerebral blood flow in man at high altitude, *Circulation Res.* **19**, 274-282 (1966).
13. C. B. Wolff, Cerebral blood flow and oxygen delivery at high altitude, *High Altitude Medicine and Biology* **1**(1), 33-38 (2000).
14. C. B. Wolff, P. Barry and D. J. Collier, Cardiovascular and respiratory adjustments at altitude sustain cerebral oxygen delivery – Severinghaus revisited, *Comp. Bioch. and Physiol. Part A* **132**, 221-229 (2002).
15. C. H. E. Imray, A. W. Wright, C. Chan, A. R. Bradwell and the Birmingham Medical Research and Expeditionary Society (BMRES), 3% carbon dioxide increases cerebral oxygen delivery when breathing hypoxic gas mixtures, *High Altitude Med. Biol.* **3**(1), p106 A31 2002 (abstract).

16. C. H. E. Imray, S. Walsh, T. Clarke, H. Hoar, T. C. Harvey, C. W. M. Chan, P. J. G. Forster and the BMRES, 3% Carbon dioxide increases cerebral oxygen delivery at 150m & 3549m, *High Altitude Med. Biol.* **3**(1), p106 A32 (2002) (abstract).
17. C. H. E. Imray, H. Hoar, A. D. Wright, A. R. Bradwell C. Chan, and the BMRES, Cerebral oxygen delivery falls with voluntary forced hyperventilation at altitude, *High Altitude Med. Biol.* **3**(1), p106 A33 (2002) (abstract).
18. C. B. Wolff, and C. H. E. Imray, Partitioning of arterial and venous volumes in the brain under hypoxic conditions, *Adv. Exp. Med. Biol.* **540,** 19-23 (2004).
19. C. B. Wolff, N. Richardson, O. Kemp, A. Kuttler, R. McMorrow, N. Hart and C. H. E. Imray, Near infra-red spectroscopy and arterial oxygen extraction at altitude, Adv. Exp. Med. Biol., **599,** 183-189.
20. R. R. Martinez, S. Setty, P. Zong, J. D. Tune and H. F. Downey, Nitric oxide contributes to right coronary vasodilatation during systemic hypoxia, *Am. J. Physiol.* **288**(3), H1139-H1146 (2005).
21. B. Folkow and E. Neil, *Circulation* (Oxford University Press, London, Toronto, 1971).
22. J. A. Guzman, A. E. Rosado and J. A. Kruse, Dopamine-1 receptor stimulation impairs intestinal oxygen utilization during critical hypoperfusion, *Am. J. Physiol.* **284**, H668-H675 (2003).
23. A. Krogh, *The Anatomy and Physiology of Capillaries* (Hafner Publishing Co., New York, 1959).
24. M. McCabe and D. J. Maguire, The measurement of the diffusion coefficient of oxygen through small volumes of viscous solution: implications for the flux of oxygen through tissues, *Adv. Exp. Med. Biol.* **316**, 467-473 (1992).
25. B. R. Duling and R. M. Berne, Longitudinal gradients in periarteriolar oxygen tension. A possible mechanism for the participation of oxygen in local regulation of blood flow, *Circulation Res.* **27**, 669-678 (1970).

NEAR INFRA-RED SPECTROSCOPY AND ARTERIAL OXYGEN EXTRACTION AT ALTITUDE

Christopher B Wolff*, Neil Richardson, Oliver Kemp, Anya Kuttler, Roger McMorrow, Nigel Hart and Christopher HE Imray**

Abstract: The ratio of oxygenated to total haemoglobin (Hb), or rSO_2, obtained by near infra-red spectroscopy (NIRS), includes both arterial and venous blood of the region examined. The relationship of arterial oxygen extraction, E, and saturation, SaO_2, to rSO_2 can be expressed, for normally functioning tissue, as $E = 1.39(1 - rSO_2/SaO_2)$. Cerebral E, at rest, is constant at lower altitudes but is reduced at 5000 m. This corresponds to constant values of E for SaO_2 values above 90% (approximately). E declines linearly for lower SaO_2 values, either including measurement at high altitude or at sea level with a reduced inspiratory oxygen concentration. In addition to measurements of brain NIRS resting oxygen extraction of liver, muscle and kidney have also been calculated from NIRS measurements made, on normal inspired air, at sea level and after acute ascent to 2400 m and 5050 m. At 5050 m E was reduced for all four regions but at 2400 m was the same as at sea level for brain, liver and muscle; for the kidney E was elevated at 2400 m. Cerebral oxygen extraction was calculated for rest and the full range of exercise. It was constant at sea level for the lower levels of exercise and, if the calculated extraction value assumptions still hold at lower SaO_2 values, reduced for the higher work rates at intermediate altitudes. The present study confirms constancy of oxygen extraction and hence the ratio of oxygen delivery to oxygen consumption (1/E), within physiological limits, and appears to show where those limits lay and, to some extent, show how matters change beyond ordinary physiological limits.

1. INTRODUCTION

Cerebral oxygen delivery is sustained at the same value as at sea level for moderate altitudes [1,2]. The constancy of MCAV (middle cerebral artery velocity) times SaO_2

* Applied Physiology, Block 9, St Thomas's Hospital, Lambeth Palace Rd., London, SE1 7EH, UK.
** Coventry and Warwickshire County Vascular Unit, University Hospitals Coventry and Warwickshire, Coventry CV2 2DH, UK

(proportional to cerebral DO_2) is illustrated in Figure 1A as a percentage of the sea level value. Figure 1 data are from Wolff and Imray [3]. A reduced DO_2 value at 5050 m and also another reduced value taken at sea level in subjects breathing 12.5% oxygen are also illustrated in Figure 1A. On the right, Figure 1B, DO_2 is plotted against SaO_2; values at SaO_2 below around 90% are the same reduced values (5050 m and 12.5% at sea level) [3]. The product of MCAV and arterial oxygen saturation (SaO_2) is also constant, at the same value as at sea level, for moderate altitudes but is reduced at 5050 m after relatively rapid ascent [3]. Since both cerebral oxygen delivery [cerebral blood flow (CBF) times arterial oxygen content (CaO_2)] and MCAV times SaO_2 agree on constancy at moderate altitudes it is assumed that MCAV times SaO_2 is proportional to, and a surrogate for, cerebral oxygen delivery.

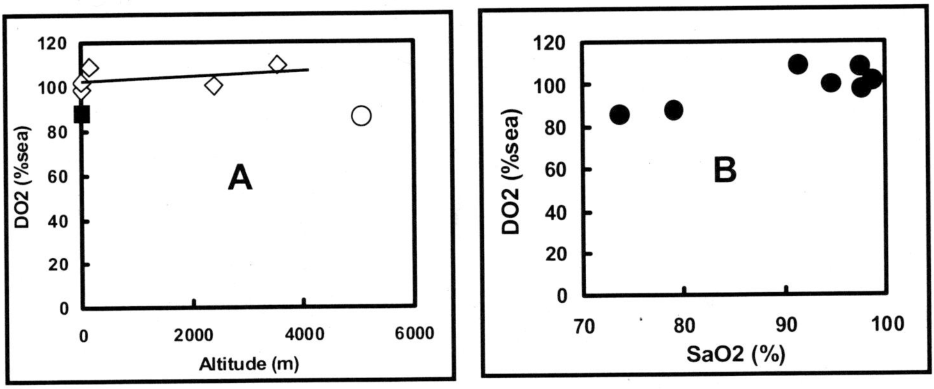

Figure 1. Cerebral oxygen delivery as a percentage of the sea level value (mean of two values). A regression line has been fitted to values at all altitudes other than 5050 m in A (◊) and its slope is not significant (equation: DO_2%sea = 102 - 0.01 x altitude). The 5050m point (○) and the one recorded at sea level on 12% O_2 (■) show reduced oxygen delivery. In B DO_2 values are plotted against SaO_2. *From Wolff, et al.* [3].

Wolff and Imray [3] showed that there was a relationship between SaO_2, the ratio (p) of arterial to venous volume in the area sampled by near infrared spectroscopy (NIRS), oxygen extraction (E) and the NIRS derived ratio of oxygenated to total blood (rSO_2) in the region examined:

$$rSO_2 = (SaO_2.p + SaO_2 (1 - E)) / (p + 1)) \quad (1)$$

2. ANALYTICAL METHODS

It was shown by Wolff and Imray [3] that p and E were unaltered at lower altitudes, recorded below 5050m, corresponding with the constancy of oxygen delivery illustrated in Figure 1A. The equation relating rSO_2, SaO_2, p and E derived by Wolff and Imray [3] can be re-arranged to give an equation for E:

$$E = (p + 1)(1- rSO_2 / SaO_2). \quad (2)$$

In the range where E and p are unchanging (lower altitudes) substitution for p can be made from the independent estimate made by McCormick, et al.[4]; they found that the arterial volume, Va, was 28% of the total blood volume, Va + Vv, where Vv is the venous volume. This gives Va/Vv = 0.39 = p. The present paper illustrates how the resulting estimate of E, $E = 1.39x(1 - rSO_2 / SaO_2)$, changes with hypoxia.

3. RESULTS

3.1. Cerebral oxygen extraction – comparison with oxygen delivery

The values for rSO_2 and SaO_2 from Wolff and Imray [3] and from a slow ascent to high altitude with partial acclimatisation (Chamlang base camp; Nepal 5000 m, Medical Expeditions - Medex - 2003) have been entered into equation 2 to obtain values of E (assuming a value for p of 0.39). E has then been plotted against altitude (Figure 2, left panel) and, also, against arterial oxygen saturation (Figure 2, right hand panel).

The constancy of the values for cerebral oxygen extraction for all recorded points breathing air below 5030 m (and with SaO_2 above 90%) correspond with cerebral DO_2 findings (Figure 1). There is constancy of E, just as for DO_2 for the lower altitudes, with corresponding higher SaO_2 values and reduced values for E for the highest altitudes (5050 and 5000 m) and lower SaO_2 values (below 90%).

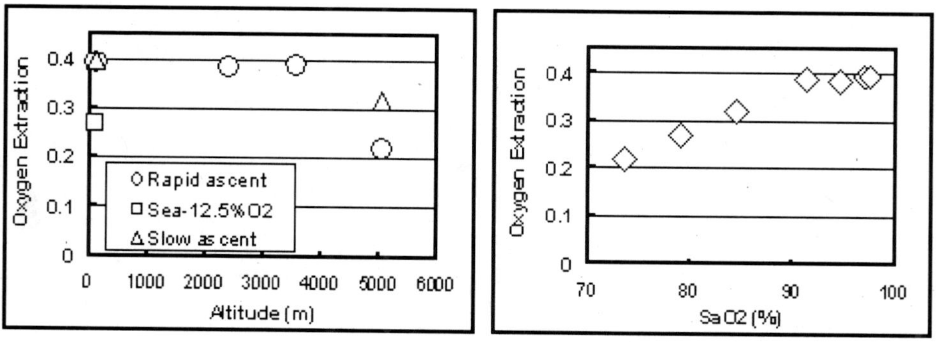

Figure 2. Cerebral oxygen extraction, calculated from $E = 1.39\ (1 - rSO_2/SaO_2)$ is shown here versus altitude on the left and against SaO_2 on the right. The points are for the same altitudes as are shown in Figure 1 (from Wolff and Imray, [3]; sea level, 2400, 3549 and 5050 m and at sea level breathing 12.5% O_2) with the addition of the mean value from a slower ascent to 5000m in 2003 (Medical Expeditions – Medex – 2003, Chamlang base camp, Nepal).

The extra value obtained from the slow ascent to Chamlang base camp at 5030 m (Nepal, 2003) shows a less severe reduction in oxygen extraction than that seen for 5050 m with a rapid ascent. Oxygen extraction values plotted against SaO_2 are constant above SaO_2 around 90%, but fall linearly for lower SaO_2 values.

3.2. Oxygen extraction of brain, liver, muscle and kidney

Values of rSO_2 and SaO_2 have been obtained from the brain, liver, resting thigh muscle and kidney at sea level, 2400 m and 5050 m (Table 1) [5]. Oxygen extractions have, again, been calculated here from $E = 1.39\ (1 - rSO_2/SaO_2)$ and appear in the lower section of

Table 1. E has been expressed as a percentage of the sea level value, E%; Figure 3 illustrates E% plotted against altitude on the left and SaO_2 on the right.

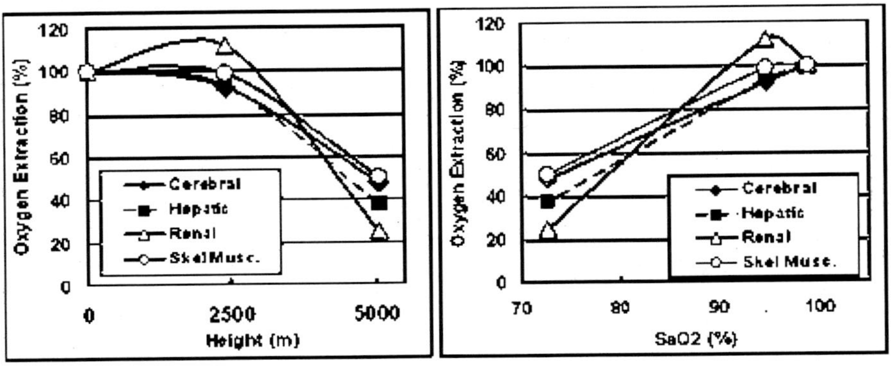

Figure 3. Oxygen extraction, expressed as a percentage of the sea level value, for brain, liver, resting skeletal muscle and kidney; derived from rSO_2 and SaO_2 obtained at sea level, 2400 m and 5050 m.[5] For brain, liver and muscle oxygen extraction was much the same at 2400 m as at sea level; but for the kidney oxygen extraction was higher. Oxygen extraction, according to the equation $E = 1.39\ (1 - rSO_2/SaO_2)$, is expressed as a percentage of the sea level value. It is reduced at 5050 m for all four organs.

Since, either E or p may have changed at the higher altitude (5000 m) and lower SaO_2 (~73%) the values calculated for E may be incorrect. However, they are unlikely to be normal. End tidal PCO_2 is also given in the table ($EtCO_2$) for comparison with other altitude studies. The end-tidal PCO_2 in the table for 5050 m, 29.4 mm Hg (acute ascent) is higher than in the study of Wolff et al. [2] (Everest base-camp), where mean PCO_2 was 26.2 mm Hg (± 0.57; SEM), arterial PO_2 43.6 mm Hg. The latter subjects were longer acclimatized probably accounting for the lower PCO_2.

Table 1. Measured values of SaO_2, End tidal PCO_2 ($EtCO_2$, mm Hg), Cerebral rSO_2 (Cer. rSO_2), Renal rSO_2 (Ren. rSO_2) and resting skeletal muscle rSO_2 (Skel. M rSO_2). [5] Measurements were made at sea level, 2400m and 5050 m. The lower section shows the values obtained for oxygen extraction, expressed in fractional form again for each organ, calculated from the equation given on the left.

	SpO_2	$EtCO_2$	Cer rSO_2	Hep rSO_2	Ren rSO_2	Sk M rSO_2
Sea	98.8	37.4	69	74.3	75.6	73.1
2400m	94.6	36.4	68.2	72.7	69.7	70.3
5050m	72.6	29.4	62.1	65.8	68.4	63.1
			Cerebral	Hepatic	Renal	Muscular
Fractional		Sea Level	0.42	0.34	0.33	0.36
E = 1.39 x (1 –		2400m	0.39	0.32	0.37	0.36
Rs/Sa)		5050m	0.20	0.13	0.08	0.18

3.3. Cerebral oxygen extraction – effects of exercise and individual variation at rest

Cerebral oxygen extraction has been calculated (using equation 2) for rest and the full range of exercise at 150 m 'sea level', 3610 m, 4750 m and 5260 m from values given in a recent study [6]. The results are illustrated in Figure 4, left panel against percentage of VO_{2Max} (maximum exercise), right panel against oxygen consumption (VO_2) and, in Figure 5A against SaO_2 (%). For 150 m E% fell progressively with increasing exercise intensity (Figure 4); for higher altitudes all showed a fall up to 50% VO_{2Max} then a rise towards the resting E value for the higher levels of exercise.

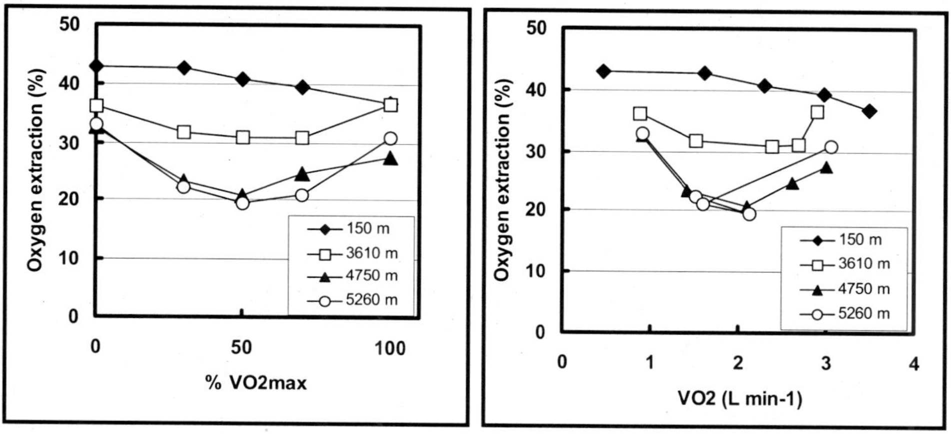

Figure 4. Cerebral oxygen extraction expressed as a percentage, for exercise between rest and VO_{2max}. The exercise studies have been undertaken at 150 m, 3610 m, 4750 m and 5260 m.[6] Each point represents the average value from 9 subjects. Values are plotted against percent of VO_{2max} on the left, VO_2 on the right.

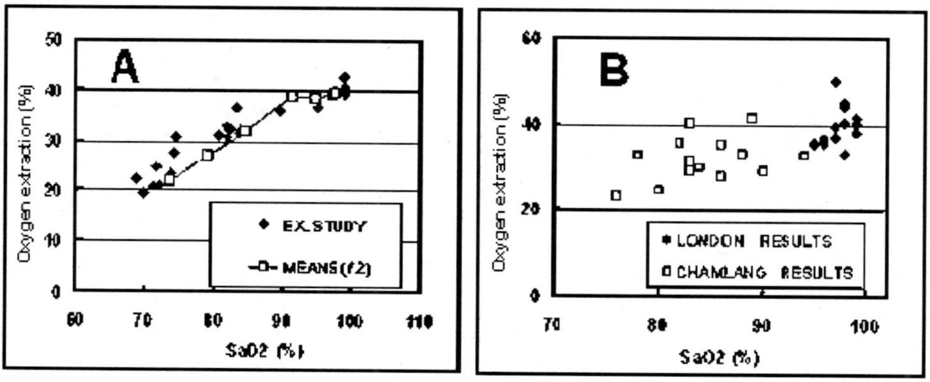

Figure 5. A. Values of cerebral E (%) from the exercise study, [6] for all altitudes and levels of exercise intensity, plotted against SaO_2 (♦) with, for comparison, mean resting values (◊) from section 4.1 above (Figure 2, right). B. Resting values of cerebral E (%) measured at sea level (London) and after slow ascent to 5030 m (Chamlang base-camp) showing variability between subjects.

The values of E in exercise are also plotted against SaO_2 in Figure 5A. The lower SaO_2 values were found at the mid range of exercise for all attitudes above sea level whereas the lowest was at VO_2 max for sea level. These low SaO_2 values in exercise correspond with the lowest values for E (Figure 5A) though they are a little higher than the trend for rest (shown in Figure 2, right hand panel) reproduced in Figure 5A (open squares).

Variations due to individual differences for resting subjects at sea level and 5030 m (slow ascent to Chamlang base-camp) are shown in Figure 5B.

4. DISCUSSION

Reductions in resting oxygen extraction (E, equation 2) were seen for brain, liver, kidney and muscle at the highest altitudes – around 5000 m. These reduced oxygen extraction values could be in error if changes occurred in p, the arterial to venous volume ratio. If true then oxygen consumption would fall too unless there was a compensatory rise in oxygen delivery, DO_2. The consistency and parallel nature of resting cerebral oxygen extraction and oxygen delivery with arterial oxygen saturations below 90% suggest a progressive pattern of deterioration of function once outside the physiological range ($SaO_2 <$ approximately 90%; Figures 1 and 2). Cerebral DO_2 reduction coupled with low oxygen extraction means a lowering of cerebral oxygen consumption.

The increased renal oxygen extraction for the modest altitude shown in Figure 3 is compatible with sustained glomerular filtration with sustained renal VO_2 in the face of the reduced arterial oxygen content. The ratio of oxygen delivery to metabolic rate is therefore reduced with a consequent increase in oxygen extraction. The resultant lowering of the renal tissue oxygen tension is the well known cause of erythropoietin secretion and hence the erythropoietic response.

The reduced cerebral oxygen extraction, E, in the middle range of exercise (Figure 4) appears largely compatible with the tendency, once outside the physiological range of SaO_2, for E to fall as saturation falls. However, the values at each arterial saturation in exercise appear to be a little higher (Figure 5) than those found in resting subjects (Figure 2, and 5A).

Figure 5B shows the range of variation of E for resting individuals both at sea level and altitude. Trends across subjects are therefore more reliable than single individual findings. This suggests individual subject control values will be especially useful where available. If no control value is available one has to see a larger deviation from mean normal values before being sure of one's interpretation of individual values.

The present study suggests that, in normal subjects, cerebral oxygen extraction, like cerebral oxygen delivery, remains constant when SaO_2 is in the normal 'physiological' range (above around 90%). Similarly for liver and resting skeletal muscle but not the kidney. All four organs show a reduction in oxygen extraction with lower SaO_2 values. The limits for sustaining oxygen delivery and extraction probably depend on oxygen content rather than just SaO_2 and will differ between major organ groups since, for example, skeletal muscle tolerates oxygen content values as low as 11 ml O_2 (100 ml blood)$^{-1}$ at 30 watt exercise [7].

This derivation of oxygen extraction from rSO_2 and SaO_2 appears to offer a useful way to assess the adequacy of tissue oxygenation and to examine the normal physiological range.

5. REFERENCES

1. C. B. Wolff, Cerebral blood flow and oxygen delivery at high altitude, *High Altitude Medicine and Biology* **1**(1), 33-38 (2000).
2. C. B. Wolff, P. Barry and D. J. Collier, Cardiovascular and respiratory adjustments at altitude sustain cerebral oxygen delivery – Severinghaus revisited, *Comp. Bioch. and Physiol. Part A* **132**, 221-229 (2002).
3. C. B. Wolff, and C. H. E. Imray, Partitioning of arterial and venous volumes in the brain under hypoxic conditions, *Adv. Exp. Med.. Biol.* **540**, 19-23 (2004).
4. P. W. McCormick, M. Stewart, M. G. Goetting and G. Balakrishnan, Regional cerebrovascular oxygen saturation measured by optical spectroscopy in humans. *Stroke* **22**, 596-602 (1991).
5. M. Beasley, C. Chan, H. Hoar, C. Knickenberg, P. Forster, C. Imray and T. And Cerebral, hepatic, renal, skeletal muscle and peripheral oxygenation at 0m, 2400m and 5050m, *High Altitude Medicine & Biology* **2**(1), p96 A71, (2001) (abstract).
6. C. H. E Imray, S. D. Myers, K. T. S. Pattinson, A. R. Bradwell, C. W. Chan, S. Harris, P. Collins and A. D. Wright, Effect of exercise on cerebral perfusion in humans at high altitude. *J. Appl. Physiol.* **99**(2), 699-706 (2005).
7. C.B. Wolff, Cardiac output, oxygen consumption and muscle oxygen delivery in submaximal exercise: normal and low O_2 states. *Adv. Exp. Med. Biol.* **510**, 279-284 (2003).

OXYGEN DELIVERY AT SEA LEVEL AND ALTITUDE (AFTER SLOW ASCENT TO 5000 METERS), AT REST AND IN MILD EXERCISE

Christopher B Wolff *, C. Douglas Thake**, Alexander Truesdell, Daniel Mattison, Lisa Handcock, David J Collier and James S Milledge[1]

Abstract: Oxygen delivery (DO_2) calculated from cardiac output, haematocrit (Hct) and arterial oxygen saturation (SaO_2), has been obtained on six subjects at sea level (London) and after slow ascent to 5000 meters (Chamlang base camp) at rest and during mild exercise (25 watts and 50 watts). Haematocrit was increased in all six subjects at 5000 m and expressed as haemoglobin (Hb) rose from a mean (+/- standard error; SEM) of 13.8 +/-0.1 g $(100\ ml)^{-1}$ to 15.8 +/-0.3 g $(100\ ml)^{-1}$ ($t = 6.3$, $p = 0.0014$). SaO_2 was almost constant with exercise at sea level (rest 98.5%, 25 w 98.3% and 50 w 98.3%) but declined more steeply with exercise at 5000 m (rest 88.8 +/-0.6%, 25 w 85.4 +/-0.4% and 50 w 84.4 +/-0.5%). Arterial oxygen content (CaO_2) was very similar for 25 watts exercise at altitude (5000 m, 18.1 ml per decilitre – dl) as at sea level (London, CaO_2 18.2 ml dl^{-1}). At rest CaO_2 was higher at altitude (18.8 +/-0.2 ml dl^{-1}) than at sea level (18.3 +/-0.4 ml dl^{-1}) and at 50 w CaO_2 was lower at altitude (17.9 +/-0.4 ml dl^{-1}) than at sea level (18.2 +/-0.2 ml dl^{-1}).

Hence, similar cardiac output values at rest (sea level, 5.0 +/-0.4 litres min^{-1} l min^{-1}; altitude, 5.6 +/-0.3 l min^{-1}) and at 25 w exercise (sea level, 8.2 +/-0.7 l min^{-1}; altitude, 8.3 +/-0.9 l min^{-1}) resulted in similar values for DO_2 at rest (sea level, 0.9 +/-0.1 l min^{-1}; altitude, 1.0 +/-0.1 l min^{-1}) and 25 w exercise (sea level, 1.5 +/-0.1 l min^{-1}; altitude, 1.5 +/-0.2 l min^{-1}). For 50 w exercise cardiac output and oxygen delivery were greater at altitude in one subject but were significantly reduced for the remaining five (cardiac output mean difference 3.0 +/-0.9 1 l min^{-1}, $p = 0.015$; DO_2 mean difference, 0.56 +/-0.2 1 l min^{-1}, $p = 0.028$). Acclimatization was therefore adequate to sustain a normal value

[1] * Applied Physiology, Block 9, St Thomas's Hospital, Lambeth Palace Rd., London, SE1 7EH.
**School of Life and Health Sciences, Coventry University, Priory St., Coventry, CV1 5FB.
***Clinical Pharmacology, William Harvey Research Institude, Barts and The London, Charterhouse Square, London, EC1M 6BQ, UK

for oxygen delivery for rest and 25 watts exercise (via compensatory erythropoiesis) but insufficient for 50-watt exercise in five of the six subjects.

1. INTRODUCTION

"....the way the human organism responds to hypoxia is important in many areas of medicine including respiratory diseases, cardiology, anaesthesia and aviation medicine." (Ward et al.[1]). The normal processes providing an adequate supply of oxygen include respiration, carriage of oxygen from the lungs to the tissues and a sufficient oxygen tension gradient between blood and tissues for the supply to sustain oxygen consumption (the oxygen cascade, Nunn[2]). The oxygen carrying power of the blood depends on the haemoglobin concentration, Hb, and cardiac output; these therefore determine the rate at which the oxygen is carried in the blood, the oxygen delivery (DO_2).

The analysis of Wolff[3], based on the experimental work of Koskolou et al.[4] and Roach et al.[5], showed that, where CaO_2 was lowered in normal subjects, the exercising muscle oxygen delivery was sustained at the expense of other tissues. The exercise in these studies was moderate and was for a limited muscle group so that there was no cardiac limitation at normal CaO_2 and yet with low CaO_2 total cardiac output was insufficient to provide normal oxygen delivery for non-exercising tissues. In chronic altitude hypoxia ventilatory acclimatization raises oxygen saturation (SaO_2) well above the level found with acute exposure and, after slow ascent, haemoglobin concentration increases.

The present study examines the effect of a slow ascent to 5000 meters on Hb, SaO_2, CaO_2 and cardiac output and hence total oxygen delivery (DO_2) in resting subjects and at 25 watts and 50 watts (mild) exercise.

2. METHODS

2.1. Experimental

Six normal subjects (4 male, 2 female) were studied, at rest and during mild exercise (25 watt and 50 watt), on a cycle ergometer (friction belt resistance type). Measurements were made at sea level (London) and at 5000 m altitude following a three-week period of ascent (Chamlang mountain basecamp in Nepal). The measurements were made after 20 minutes rest and at 3 to 5 minutes after the onset of exercise (steady state). Non-invasive arterial blood pressure (finger cuff; Finapres, model 2300, Ohmeda, Madison Wisconsin, USA) and SaO_2 (pulse oximeter; Novametrix Medical Systems, Inc. Wallingford, CT, USA; model 515A) were recorded via an A-D converter (micro 1401, Cambridge Electronic Design – CED), using CED software (Spike2).

Ventilation was measured by dry gas meter (Harvard Apparatus Ltd., Edenbridge, Kent, UK) from expired gas collected in Douglas bags at rest and during the 3-5 minutes of exercise. Expired oxygen was measured using a paramagnetic gas analyser (Capnograph, Morgan Medical Ltd., Gillingham, UK). Temperature and barometric pressure at sea level were read from a mercury-in-glass barometer (F. Darton and Co. Ltd., Watford, UK). Barometric pressure according to a Suunto wristwatch (Observer model, Helsinki, Finland) was compared with the mercury-in-glass barometer at sea level and used to obtain barometric pressure at 5000 m.

Haematocrit was measured by means of a micro-centrifuge (at sea level, Hawksley & Sons Ltd. Lancing, UK and at 5000 m, Clinispin Hct Benchtop Centrifuge, Phitec, Milton Keynes, UK).

2.2. Analytical

Cardiac output was derived from arterial blood pressure (BP) using the algorithm of PulseCO (LiDCO, London, UK). Arterial volume is derived, V=250(1-e^-(0.0092xBP)), the mean value subtracted and auto-covariance calculated on the result. The duration to the first positive peak is the cardiac period and stroke volume the mean of the square roots of the value at the origin (RMS) and of the absolute value of the first auto-covariance trough. The calibration was provided by calculating DO_2 at rest with the uncalibrated cardiac output then scaling it up to make DO_2 equal to 3.6 times the rate of oxygen consumption (VO_2) in the resting subjects at sea level (examples from the literature – DO_2 1.2 l min^{-1}. and VO_2 0.35 l min^{-1} gives a ratio of 3.43 [3]; and using a DO_2 of 0.96 l min^{-1} with a VO_2 of 0.25 l min^{-1} gives a ratio of 3.84 [7]).

DO_2 = arterial oxygen content (CaO_2) times cardiac output and CaO_2 = Hb x 1.34 x SaO_2 + 0.003 x PO_2 (mm Hg). Hb was calculated from Hct: Hb = 0.33 x Hct [6]. The addition of 0.003 x PO_2 makes a small difference, so error is small when using a PO_2 value derived from SaO_2.

3. RESULTS

The haematocrit (Hct) was increased at 5000 m (Chamlang base camp) in all six subjects (Figure 1 – left hand panel). At sea level arterial oxygen saturation (SaO_2) decreased only slightly with exercise. At 5000 m SaO_2 was lower than at sea level and fell much more with exercise (Figure 1 – right hand panel).

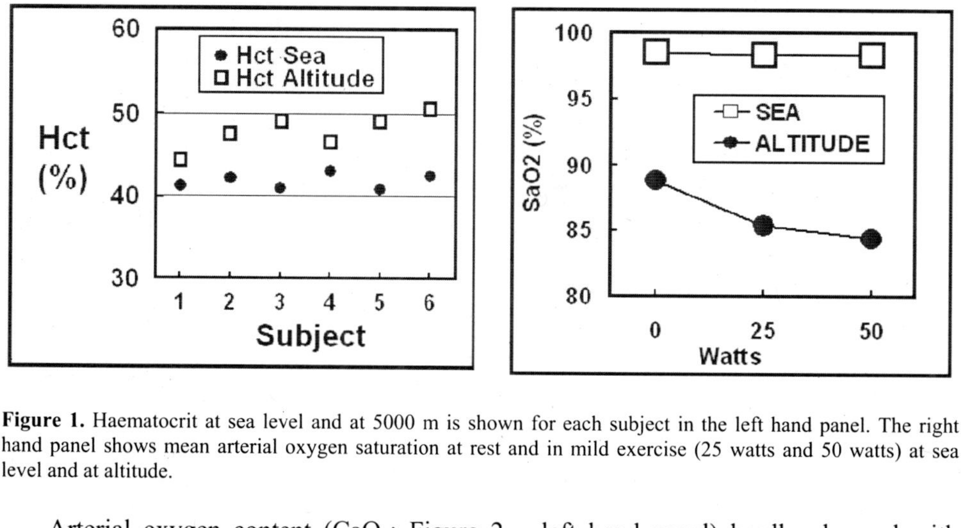

Figure 1. Haematocrit at sea level and at 5000 m is shown for each subject in the left hand panel. The right hand panel shows mean arterial oxygen saturation at rest and in mild exercise (25 watts and 50 watts) at sea level and at altitude.

Arterial oxygen content (CaO_2; Figure 2 – left hand panel) hardly changed with exercise at sea level (18.3 +/-0.2 ml dl^{-1} at rest and 18.2 +/-0.2 ml dl^{-1} at both levels of exercise) but at altitude CaO_2 fell with exercise from 18.8 +/-0.4 ml dl^{-1} (higher than at sea level), to much the same as the sea level value at 25 watts exercise (i.e. 18.1 +/-0.4 ml dl^{-1}) and a lower value for 50 watts exercise (17.9 +/- 0.4 ml dl^{-1}).

Resting oxygen consumption (VO_2) at sea level has been plotted for each subject against body weight (Figure 2 – right hand panel). The thin line is a least squares regression fit (VO_{2STP} at rest = 0.0038Wt - 0.005, where VO_2 is in litres min^{-1} and weight is in kilograms). The slope of the line through the origin is virtually identical (VO_{2STP} at rest = 0.0038Wt, bold regression line).

The non-scaled sea level value for resting cardiac output (algorithm from LiDCO™, UK) for each subject is presented in Table 1 with an initial estimate of oxygen delivery (non-scaled cardiac output times CaO_2). This allows calculation of the correction (scaling) factor for cardiac output. Table 1 shows this scaling factor which, on application to the raw cardiac output, results in a resting DO_2 (at sea level) of 3.6 times VO_2.

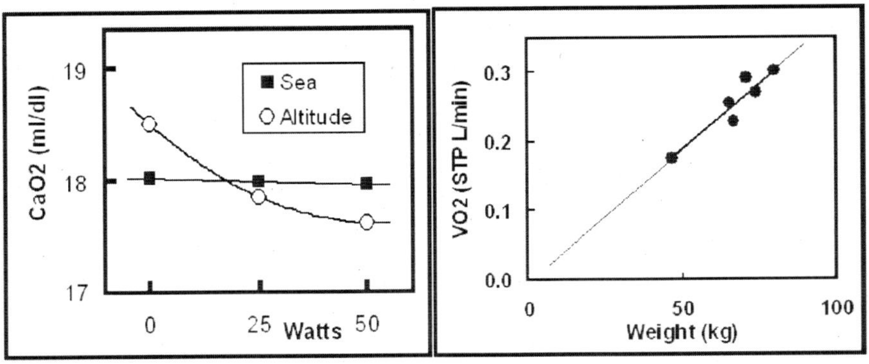

Figure 2. On the left, CaO_2 is plotted against watts for both sea level and 5000 m altitude. There is a close similarity between values at 25 watts, while the 5000 m value for rest is higher and the value for 50 w exercise is lower than the sea level value. On the right individual resting VO_2 values have been plotted against body weight (London). VO_2 (ml STP per minute) = 0.0038 x body weight (in kg).

Table 1. DO_2 raw (non-scaled) in the table is the DO_2 value obtained by multiplying the raw (non-scaled) resting cardiac output by arterial oxygen content (CaO_2). The scaling factor for cardiac output is then obtained as the value needed to raise the raw DO_2 to 3.6 times VO_2 (See text). VO_2 is in ml. min^{-1} at standard temperature and pressure.

SUBJECT	DO_2 raw	VO_2 Sea, rest	DO_2:VO_2 ratio	Scaling factor
1	704	229	3.08	1.17
2	607	302	2.01	1.79
3	293	176	1.67	2.16
4	472	255	1.85	1.95
5	626	271	2.31	1.56
6	405	291	1.39	2.59

Cardiac output (raw value times scaling factor), CaO_2 and DO_2 are given in Table 2 for rest and for 25 and 50 w exercise; values are listed both for sea level (London) and 5000 m (Chamlang) following a conservative ascent profile. All subjects had been resident at Chamlang for at least two days.

Mean cardiac output has been plotted in Figure 3 (left hand panel). Mean cardiac output for rest and 25 w exercise were much the same at sea level (rest 5.0 +/-0.4 l min^{-1}, 25 w 8.2 +/-0.7 min^{-1}) and at 5000 m (rest 5.6 +/-0.3 l min^{-1}, 25 w 8.3 +/-0.9 min^{-1}), but cardiac output was reduced at altitude for 50 w exercise (sea level 12.1 +/-1.4 min^{-1}; 5000 m 9.9 +/-1.1 min^{-1}). The reduction was not significant across the six subjects, but for the five subjects where there was a reduction the mean value at altitude (10.0 +/-1.4 min^{-1}) was significantly less than at sea level (13.0 +/-1.2 min^{-1}; difference 3.0 +/-0.9 min^{-1} paired t test, $p < 0.03$).

Mean DO_2 values were, for sea level and altitude respectively, rest 0.91 +/-0.07 l min^{-1} and 1.05 +/-0.06 l min^{-1}, 25 w exercise 1.49 +/-0.13 l min^{-1}, 1.50 +/-0.17 l min^{-1} and for 50 w exercise 2.20 +/-0.25 l min^{-1}, 1.77 +/-0.21 l min^{-1}. Again, for five subjects (2-6) there was a significant reduction in DO_2 for 50 w at altitude (from 2.27 l min^{-1} to 1.82 l min^{-1}, $p < 0.03$, paired t test).

The relationship between VO_2 and watts obtained here was $VO_2 = 0.0173\text{watts} + 0.2786$ where VO_2 is in l min^{-1} and mean values were 0.254 +/-0.019, 0.759 +/-0.080 and 1.117 +/-0.124 l min^{-1} for rest, 25 watts and 50 watts.

Table 2. Values here are for cardiac output (CO), CaO_2 and DO_2 at sea level (London) and 5000 m (Chamlang base camp) after a slow ascent (conservative ascent profile).

Subject	WATTS	CO	CO	CaO_2	CaO_2	DO_2	DO_2
Sex		Sea	Altitude	Sea	Altitude	Sea	Altitude
1	0	4.6	6	18.1	17.3	823	1037
M	25	5.9	8.6	18.1	16.6	1069	1430
	50	7.3	9.3	18.1	16.5	1316	1530
2	0	5.9	6.7	18.5	18.6	1088	1253
M	25	9.2	10.2	18.3	17.9	1689	1825
	50	16.8	13	18.3	17.6	3071	2277
3	0	3.5	5.2	18	19.3	632	995
F	25	9.1	9.5	18	18.5	1650	1753
	50	14.7	11.4	18	18.1	2643	2066
4	0	4.8	4.6	19	18	919	819
F	25	6.8	3.9	18.9	17.4	1293	685
	50	10.8	4.8	18.9	17.2	2032	832
5	0	5.5	6	17.7	19.8	976	1187
M	25	7.6	8.4	17.7	18.9	1340	1578
	50	10.5	9.7	17.7	18.7	1859	1813
6	0	5.7	5.1	18.3	19.8	1047	1001
M	25	10.4	9.1	18.3	19.1	1904	1750
	50	12.3	11	18.3	19	2261	2091

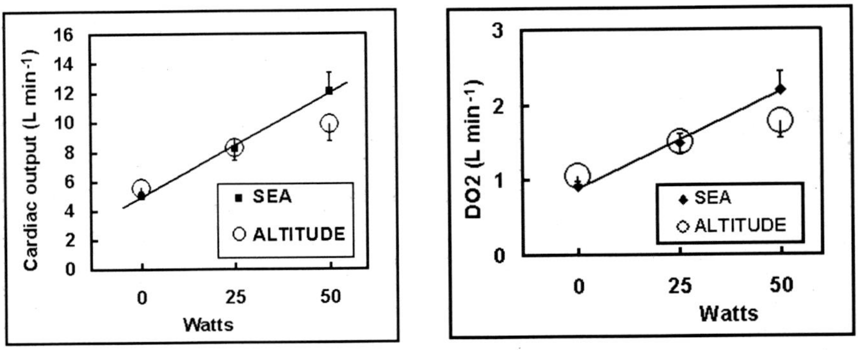

Figure 3. Left panel, cardiac output (CO). Mean CO is virtually the same for rest and 25-watt exercise at 5000 m as at sea level. However, CO is reduced for 50 w exercise at 5000 m (significantly so in 5 subjects). Right panel, oxygen delivery, DO_2, shows a reduced mean value at 5000 m (Altitude) compared with sea level; again, DO_2 was significantly reduced in 5 of the six subjects.

4. DISCUSSION

The finding that oxygen delivery (DO_2) after a slow ascent to 5000 m is sustained for rest and for 25 watts exercise is consistent with adequate acclimatization for routine activity (average metabolic rate). The finding that there was a significantly lower DO_2 at 50 w in five out of the six subjects is consistent with there having been a limitation of cardiovascular adjustment with more than very modest exercise. The physiological difficulty faced here is that one oxygen content value (CaO_2) cannot cover a wide range of exercise, since there is a fall in SaO_2 with increasing exercise, at least over the lower levels of exercise, due to reduced capillary transit time; there is diffusion limitation.

Cardiac output measurement from non-invasive blood pressure is dependent on the algorithm developed by LiDCO (London). In clinical use it is calibrated from an orthodox, one shot, lithium dilution measurement. Here, the subjects are normal and at rest during the sea level control measurements. The method of scaling the derived cardiac output measurement utilized here depends on the normal relationship between oxygen delivery and oxygen consumption. The fact that the resting oxygen consumption measurements at sea level plotted well against the subjects' weights (Figure 2 – right hand panel) allowed the use of a scaling factor to scale the calculated DO_2 value to be 3.6 times VO_2. This gives a potential margin of error, though there is some support for it from the linear relationship between the resultant cardiac outputs at sea level and the work rate[3]. The algorithm appears reasonably robust [8].

The present study suggests there is good cardiovascular acclimatization after a slow ascent to 5000 m, at least for rest and 25 watts exercise, but that for 50-watt exercise cardiac output and oxygen delivery were compromised in 5 out of six subjects.

5. ACKNOWLEDGEMENTS

This work was carried out in association with Medical Expeditions (Altitude Education and Research Charity). Thanks are due to Ted Carter, Physiological Laboratories, Queen Mary College, London, for technical assistance.

6. REFERENCES

1. Ward, M. P., J. S. Milledge, and J. B. West, High Altitude Medicine and Physiology, Chapman and Hall Medical, London (1989)
2. J. Nunn, Applied Respiratory Physiology: With Special Reference to Anaesthesia, Butterworths, London (1971)
3. C.B. Wolff, Cardiac output, oxygen consumption and muscle oxygen delivery in submaximal exercise: normal and low O_2 states. Adv. Exp. Med. Biol. **510**: 279-284, (2003)
4. M. D Koskolou, , R. C. Roach, J. A. Calbet, G. Rådegran and B. Saltin, Cardiovascular responses to dynamic exercise with acute anemia in humans. Am. J. Physiol. (Heart and Circulatory Physiology) **273**: H1787-H1793 (1997).
5. R. C. Roach, M. D. Koskolou, J. A. L. Calbet and B. Saltin, Arterial O_2 content and tension in regulation of cardiac output and leg blood flow during exercise in humans. Am. J. Physiol. (Heart and Circulatory Physiology) **276**: H438-H445 (1999).
6. F. Habibzadeh, M. Yadollahie, M. Roshanipoor and M. Haghshenas Derivation of blood hemoglobin concentration from hematocrit: a simple method for rural areas. M Arch. Irn. Med. **4**(3): 120-122 (2001)
7. P-O. Astrand, and K. Rodahl, Textbook of Work Physiology: Physiological Basis of Exercise, 3rd Edition, McGraw-Hill International Editions, Medical Science Series, New York (1986).
8. A. Rhodes and R. Sutherland Arterial pulse power analysis: the LiDCO™ Plus system.In, Functional Haemodynamic Monitoring Update in Intensive Care and Emergency Medicine 42. Eds., Pinsky MR and Payen D, Springer-Verlag, Berlin, Heidelberg, pp183-192 (2005).

INCREASED SENSITIVITY TO TRANSIENT GLOBAL ISCHEMIA IN AGING RAT BRAIN

Kui Xu, Xiaoyan Sun, Michelle A. Puchowicz, and Joseph C. LaManna*

Abstract: Transient global brain ischemia induced by cardiac arrest and resuscitation (CAR) results in reperfusion injury associated with oxidative stress. Oxidative stress is known to produce delayed selective neuronal cell loss and impairment of brainstem function, leading to post-resuscitation mortality. Levels of 4-hydroxy-2-nonenal (HNE) modified protein adducts, a marker of oxidative stress, was found to be elevated after CAR in rat brain. In this study we investigated the effects of an antioxidant, *alpha*-phenyl-*tert*-butyl-nitrone (PBN) on the recovery following CAR in the aged rat brain. Male Fischer 344 rats (6, 12 and 24-month old) underwent 7-minute cardiac arrest before resuscitation. Brainstem function was assessed by hypoxic ventilatory response (HVR) and HNE-adducts were measured by western blot analysis. Our data showed that in the 24-month old rats, overall survival rate, hippocampal CA1 neuronal counts and HVR were significantly reduced compared to the younger rats. With PBN treatment, the recovery was improved in the aged rat brain, which was consistent with reduced HNE adducts in brain following CAR. Our data suggest that aged rats are more vulnerable to oxidative stress insult and treatment with PBN improves the outcome following reperfusion injury. The mechanism of action is most likely through the scavenging of reactive oxygen species resulting in reduced lipid peroxidation.

1. INTRODUCTION

Transient global brain ischemia induced by cardiac arrest and resuscitation results in reperfusion injury associated with oxidative stress which produces delayed selective neuronal cell loss[1,2]. With aging, increased susceptibility to oxidative stress leads to increased incidence of cardiac arrest and stroke. The age-related changes in the brain could alter the outcome of diseases, as well as the approach to therapeutic strategies.

* Kui Xu, Xiaoyan Sun, Michelle A. Puchowicz and Joseph C. LaManna, Case Western Reserve University, Cleveland, Ohio, 44106 USA

4-hydroxy-2-nonenal, a reactive product of lipid peroxidation, is considered to be cytotoxic and can be used as a marker of oxidative stress[3]. HNE reacts with proteins leading to enzyme inactivation and subsequence cellular dysfunction[4,5]. Levels of HNE-modified protein adducts have been shown to be elevated in brain during the first 24 hours of recovery following CAR[6]. Reperfusion injury can lead to secondary challenge in brainstem, which contains respiratory and cardiovascular centers. Post-resuscitation mortality is associated with impairment of brainstem function. Hypoxic ventilatory response reflects brainstem function and can be used to evaluate brainstem damage. Various antioxidants have been reported to attenuate selective ischemic hippocampal CA1 vulnerability, including a spin-trapping agent *alpha*-phenyl-*tert*-butyl-nitrone[7]. We hypothesized that an antioxidant treatment with PBN in the aged rats will result in improved outcome following CAR by reducing oxidative stress through the attenuation of lipid peroxidation. In this study we investigated the effects of PBN in a rat model of CAR and its potential as a treatment strategy to improve the outcome in aged rats. We measured physiological variables, overall survival rate, hippocampal CA1 neuronal counts and HVR in the long-term recovery rats at 6, 12, and 24 months, and 24 months treated with PBN post CAR. We also used western blot analysis to detect HNE adducts in whole tissue cortical and brainstem lysates to assess oxidative stress.

2. METHODS AND MATERIALS

2.1. Animal preparation

Fischer 344 rats (6, 12 and 24-months old) were purchased and allowed to acclimate in the animal facility at Case Western Reserve University for one week before being utilized. Surgical procedures for each experiment were as follows: anesthesia is induced by isoflurane (2.5% isoflurane, 70% N_2O in O_2) and maintained with 1-2% isoflurane, 70% N_2O in O_2 through a nasal cone. Cannulae were placed in: (i) Ventral tail artery using polyethylene tubing (PE-50, 0.023" i.d., 0.038" o.d.) for the purpose of monitoring of systemic arterial blood pressure and to obtain samples for blood gas, plasma glucose and lactate determinations (ii) External jugular vein into the right atrium using a Silastic catheter (0.025" i.d., 0.047" o.d.) for administration of drug. After surgery, the rats were allowed to recover for at least 1 hour while restrained in plastic cages. Throughout the experiment, the body temperature was maintained at 37°C by an infrared heat lamp (250W, 45 cm above the body) regulated by feedback from a rectal probe.

2.2. Induction of total cerebral ischemia in rat

Reversible total cerebral ischemia was achieved using a CAR model[2,8]. Cardiac arrest was induced in the conscious rat by rapid sequential intra-atrial injection of d-tubocurare (0.3mg) and ice-cold KCl solution (0.5M; 0.12 ml/100g of body weight). Resuscitation was initiated 5 min. after arrest following orotreacheal intubation with a 14-gauge catheter attached to a rodent ventilator (100% O_2, tidal volume: 10cc/kg, respiratory rate: 80 breaths/min). Simultaneous chest compressions and the infusion of normal saline (0.5 ml/min) were given until a spontaneous heartbeat returned. Epinephrine (4-10g) was given intravenously to establish a mean blood pressure greater than 80% of the pre-arrest value, at which point the animal was considered to be resuscitated (~7 min ischemia).

Ventilation was then adjusted to ~30% O_2 and 70% N_2O, depending on the normal range of blood gas, until spontaneous respiration was regained. For the PBN-treated rats, PBN (100 mg/kg) was infused intravenously immediately after resuscitation for 60 min. The untreated rats were given normal saline for same period of time. Non-arrested rats went through the same surgical procedures except cardiac arrest. For overall survival and hippocampal neuronal survival determinations, rats were allowed to survive for 4 days after CAR. For western blotting analysis, rats were decapitated 1 hour after resuscitation; samples of cortex and brainstem were dissected and frozen in liquid nitrogen and stored at -80°C.

2.3. Hippocampal neuron counts

Rats surviving 4 days and non-arrested controls were deeply anesthetized with isoflurane, perfused through the heart with about 200 ml 0.1 M PBS, and perfusion fixed with 4% paraformaldehyde in 0.1 M PBS (pH 7.4). The brains were removed and embedded with paraffin and sectioned on a microtome. Neuronal cell counts were made from H & E stained 5 μm sections through anterior hippocampus. At the level of atlas plate 30[9] the entire length of the pyramidal cell layer in the hippocampus was viewed under a high-power light microscope (400x). Neurons with rounded cell bodies and clearly visible nucleoli were considered to have survived. The number of neurons surviving was evaluated in CA1 region of hippocampus.

2.4. Western Blot Analyses

Whole tissue lysate was prepared and assayed for HNE modified proteins as preiously described[6]. Cortical and brainstem samples were homogenenized in ice-cold buffer (20mM HEPES, pH 7.5, 1.5 mM $MgCl_2$, 0.2mM EDTA, 0.1M NaCl) containing 0.2mM dithiothreitol and protease inhibitors (0.4mM phenylmethylsulfonyl fluoride and 2μg/ml each of leupeptin, pepstatin and aprotinin). NaCl was added to a final concentration of 0.45M and homogenate was centrifuged at 10,000G for 30 min. The supernatant was collected and stored at -80°C. Samples containing 100 μg of protein were electrophoresed on 10% SDS-polyacrilamide gels. The proteins on the gels were transferred to nitrocellulose membranes then incubated with 5% skim milk in Tris-buffered saline for 1 hour (room temperature) to block nonspecific binding. HNE modified proteins were detected by incubating the membranes with a 1:1000 dilution of polyclonal anti-HNE antibody (Calbiochem) overnight (4°C) followed by incubation with horseradish peroxidase-conjungated anti-rabbit IgG (1:5000) for 1 hour (Jackson ImmunoResearch). The primary antibody immunoreactive protein bands were visualized using enhanced chemiluminescence detection system (ECL kit, Amersham).

2.5. Statistical Methods

All values were represented as mean ± S.D. Statistical analyses were performed using SPSS v13.0 for Windows. Group comparisons are made by one-way analysis of variance (ANOVA) using Tukey's statistic or t-test. Significance was considered at the level of $p < 0.05$.

3. RESULTS

3.1. Physiological variables

As seen in Table 1, there are no significant differences in any of the physiological variables among any of the experimental groups. Blood gases were taken 1 hour after resuscitation for adjustment of ventilation to achieve PO_2 and PCO_2 values in the normal range. Typically, the rats regained spontaneous respiration 2 to 6 hours after resuscitation.

Table 1. Physiological variables (mean ± S.D.) of the long-term recovery rats.

Variable	6 mos (n =8)	12 mos (n = 5)	24 mos (n =12)	24 mos + PBN (n = 8)
Body weight (g)	389 ± 14	438 ± 44	405 33	372 ± 21
MABP (mmHg)	119 ± 2	116 ± 7	113 ± 6	119 ± 4
Pre 1 hr	113 ± 10	101 ± 16	105 ± 20	119 ± 8
Arterial pH (unit)	7.43 ± 0.04	7.41 ± 0.04	7.43 ± 0.01	7.45 ± 0.05
Pre 1 hr	7.43 ± 0.06	7.39 ± 0.06	7.44 ± 0.03	7.42 ± 0.06
PaO_2 (mmHg)	91 ± 12	92 ± 13	91 ± 15	99 ± 18
Pre 1 hr	102 ± 16	89 ± 16	96 ± 13	96 ± 11
$PaCO_2$ (mmHg)	42 ± 5	42 ± 2	43 ± 4	40 ± 4
Pre 1 hr	36 ± 3*	38 ± 2*	36 ± 3*	39 ± 5
Hematocrit (%)	49 ± 2	50 ± 1	48 ± 4	48 ± 2
Pre 1 hr	49 ± 2	49 ± 1	48 ± 3	48 ± 3
Glucose $_{plasma}$ (mM)	7.6 ± 0.6	8.0 ± 1.1	7.8 ± 2.0	7.5 ± 1.6
Pre 1 hr	8.5 ± 1.9	10.0* ± 0.7	9.6 ± 3.0	9.1 ± 1.5
Lactate $_{plasma}$ (mM)	1.5 ± 0.6	1.6 ± 0.1	1.4 ± 0.2	1.4 ± 0.4
Pre 1 hr	2.7 ± 1.2	2.4 ± 0.2*	3.9 ± 2.0*	2.5 ± 0.8

* indicates the post-resuscitation value was significantly different (t-test, $p<0.05$) from the pre-arrest value in the same experimental group.

3.2. Overall survival rate

As seen in Table 2, less than half of the 24-month old rats survived for 4 days after 7-minute CAR; with a survival rate similar to that of 3-month old adult rats with 12-minute cardiac arrest as reported before[8]. The data showed that with aging the 4 day overall survival rates were decreased, however, with the treatment by PBN the overall survival rate was improved.

Table 2. Survival (4 days) of aging rats after cardiac arrest and resuscitation.

Experimental groups	Number of deaths				Survival rate
	< 1 day	1-2 days	2-3 days	3-4days	4 days/total
6 month	0	1	0	1	75% (6/8)
12 month	2	0	0	0	60% (3/5)
24 month	1	5	1	0	42% (5/12)
24 month + PBN	0	0	0	3	63% (5/8)

3.3 Hippocampal neuronal counts

The non-arrested hippocampal neuronal counts were similar among all age groups (6 month: 116 ± 1; 12 month: 114 ± 2; 24 month: 117 ± 5; n =3 each). Hippocampal neuronal loss was shown in all age groups 4 days after resuscitation and the hippocampal CA1 counts were decreased with aging (Figure 1). The cell counts in the 6-month old rats (72 ± 30, n = 6) were significantly higher ($p < 0.05$) than that in the 24-month rats (18 ± 6, n = 5). The survival of CA1 neurons after CAR in 12-month old rats (63 ± 21, n =3) was intermediate to the 6 and 24 month old rats. In the 24-month old rats, at 4 days recovery, the PBN-treated group had significantly ($p < 0.05$) improved hippocampal CA1 cell counts (59 ± 26, n = 4) compared to the untreated rats (18 ± 6, n = 5).

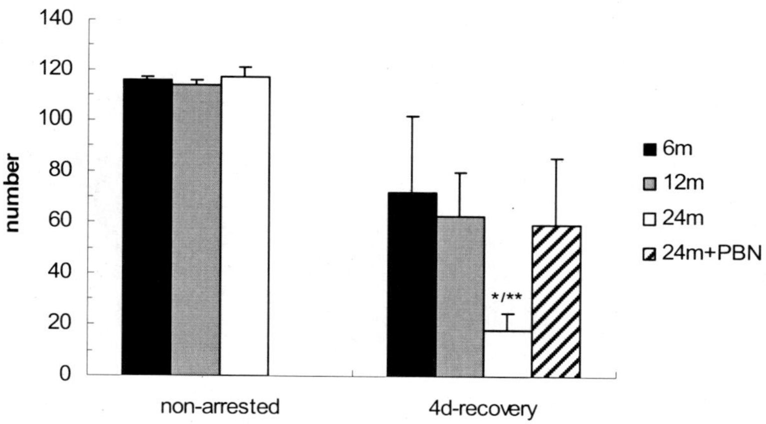

Figure 1. Hippocampal CA1 neuronal counts 4 days after cardiac arrest and resuscitation. 6m, 12m, 24m : 6, 12, and 24 month old untreated groups. 24m + PBN: 24-month old with PBN treatment. * indicates significant difference (ANOVA, $p < 0.05$) from the non-arrested group with same age. ** indicates significant difference (ANOVA, $p < 0.05$) from the PBN-treated group.

3.4. Hypoxic ventilatory response

Brainstem function was assessed by measuring the hypoxic ventilatory response. When exposed to hypoxic gas (in this case 10% O_2), mammals respond with an increase in respiration frequency and tidal volume, resulting in an increase in the minute volume. The ratio of the hypoxic minute volume to the normoxic minute volume is the so-called hypoxic ventilatory response, or HVR. The magnitude of the HVR is inversely proportional to the hemoglobin saturation[10]. Our results indicated that HVR decreases with aging as there was a significant ($p < 0.05$) decrease (~30 %) in HVR in 24-month old rats compared to the 6-month old rats (Figure 2). After CAR, the HVR was further decreased presumably because of impairment of brainstem function. Nevertheless, PBN-treated 24-month old rats had significantly higher ($p < 0.05$) HVR at 1 and 2 days of recovery after CAR (Figure 3).

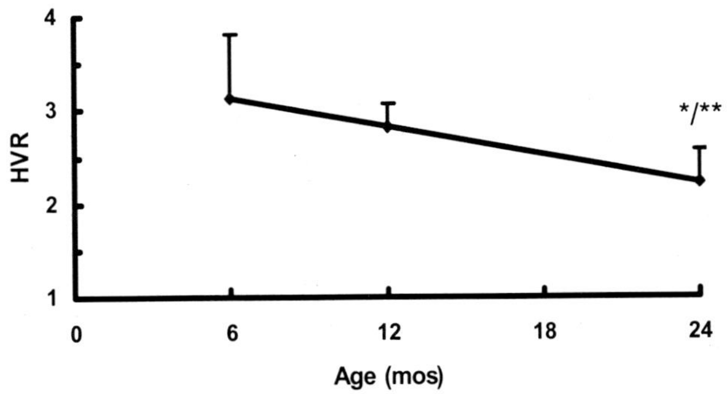

Figure 2. Hypoxic ventilatory response in aging rats. 6 mos (3.1 ± 0.7, n = 12), 12 mos (2.8 ± 0.3, n =6), 24 mos (2.2 ± 0.3, n = 20). * indicates significant difference (t-test, $p < 0.05$) from the 6-month old group; ** indicates significant difference (t-test, $p < 0.05$) from the 12-month old group.

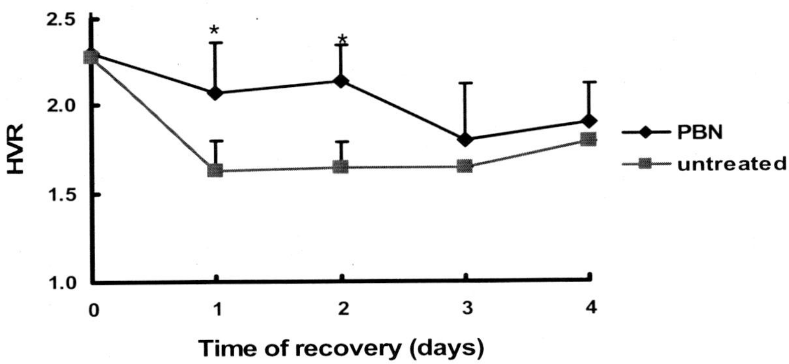

Figure 3. HVR in 24-month old rats (unteated and PBN-treated) after cardiac arrest and resuscitation, represented by ratio of minute volume, hypoxic (10% oxygen) vs. normoxic baseline. * indicates significant difference (t-test, $p < 0.05$) from the untreated group.

3.5. HNE detection

Western blot analysis showed HNE-modified proteins in the cortex and brainstem (whole tissue) of 24-month old rats. HNE adduct formation was observed within the molecular weight range of 65 to 90 kDa. Compared to the non-arrested control, the production of HNE adducts was elevated in the untreated 1 hour recovery sample. With the treatment by PBN, HNE adduct levels were reduced and similar to that of the non-arrested controls (Figure 4).

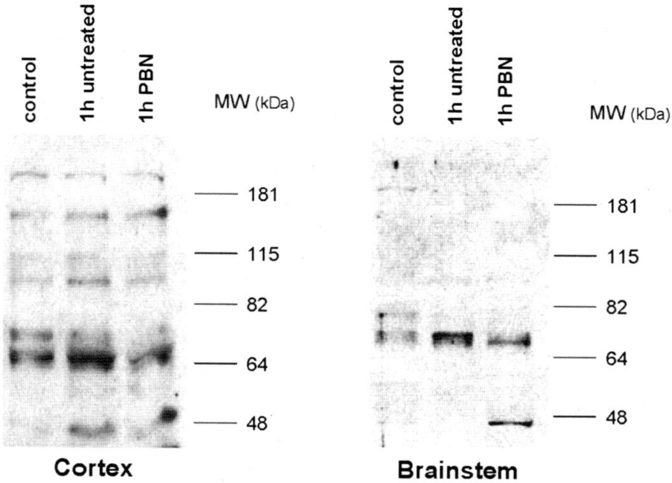

Figure 4. Western blot analyses showing HNE adducts formation in the cortex and brainstem of 24-month old rats. Control: non-arrested control; 1h untreated: untreated rat at 1 hour recovery; 1h PBN: PBN-treated rat at 1 hour recovery.

4. DISCUSSION

In rats, resuscitation from cardiac arrest manifests high mortality and delayed neuronal death in the survivors. Oxidative stress plays an important role in ischemia/reperfusion injury induced by cardiac arrest and resuscitation; with aging, the sensitivity to oxidative damage is increased. In our rat model of cardiac arrest and resuscitation we have shown that 3-month old adult rats with 12-minute cardiac arrest have similar mortality (~50%) and hippocampal neuronal death[8] compared to that of 24-month aged old rats with 7-minute cardiac arrest. Our data showed that in the 24-month old rats, both survival rate and hippocampal CA1 neuron counts were significantly reduced compared to the 6 and 12-month rats and with PBN treatment, the outcome was considerably improved. These results are consistent with the improvement of HVR with PBN treatment. Our data also showed in aging brain that the elevated HNE adducts following cardiac arrest and resuscitation were reduced with PBN treatment, suggesting that PBN's mechanism of action is through the scavenging of reactive oxygen species resulting in reduced lipid peroxidation.

In conclusion, these data suggest that aged rats are more vulnerable to reperfusion injury following cardiac arrest and resuscitation. Treatment with PBN improves the outcome following oxidative stress insult, supporting the need to explore the protective properties of antioxidants as a therapeutic approach in aging for improving outcomes from cardiac arrest and resuscitation.

5. ACKNOWLEGEMENTS

This work was supported by NIH grants NS 46074 and GM 066309. We would like to especially thank Andre Ivy for his assistance in preparation of the manuscript.

6. REFERENCES

1. J. C. Chavez and J.C. LaManna: Activation of hypoxia inducible factor-1 in the rat cerebral cortex after transient global ischemia: potential role of insulin like growth factor-1. *J.Neurosci.* **22**:8922-8931 (2002).
2. R. C. Crumrine and J.C. LaManna: Regional cerebral metabolites, blood flow, plasma volume and mean transit time in total cerebral ischemia in the rat. *J.Cereb.Blood Flow Metab.* **11**:272-282 (1991).
3. E. D. Hall, P.K. Andrus, J.S. Althaus, and P.F. Von Voigtlander: Hydroxyl radical production and lipid peroxidation parallels selective post-ischemic vulnerability in gerbil brain. *J.Neurosci.Res.* **34**:107-112 (1993).
4. K. M. Humphries and L.I. Szweda: Selective inactivation of alpha-ketoglutarate dehydrogenase and pyruvate dehydrogenase: reaction of lipoic acid with 4-hydroxy-2- nonenal. *Biochem.* **37**:15835-15841 (1998).
5. D. T. Lucas and L.I. Szweda: Cardiac reperfusion injury: aging, lipid peroxidation, and mitochondrial dysfunction. *Proc Natl.Acad.Sci U.S A* **95**:510-514 (1998).
6. LaManna JC, NL Neubauer, and JC Chávez: Formation of 4-hydroxy-2-nonenal-modified proteins in the rat brain following transient global ischemia induced by cardiac arrest and resuscitation. In: "Maturation Phenomenon in Cerebral Ischemia IV," Bazan NG et al.., eds., Springer-Verlag, Berlin, pp. 223-227 (2001).
7. J. W. Phillis and C. Clough-Helfman: Protection from cerebral ischemic injury in gerbils with the spin trap agent N-tert-butyl-alpha-phenylnitrone (PBN). *Neurosci.Lett.* **116**:315-319 (1990).
8. J. M. Hoxworth, K. Xu, Y. Zhou, W.D. Lust, and J.C. LaManna: Cerebral metabolic profile, selective neuronal loss, and survival of acute and chronic hyperglycemic rats following cardiac arrest and resuscitation. *Brain Res.* **821**:467-479 (1999).
9. Palkovits M and MJ Brownstein: Maps and Guide to Microdissection of the Rat Brain. Elsevier, New York, (1988).
10. J. V. Weil, E. Byrne-Quinn, I.E. Sodal, W.O. Friesen, B. Underhill, G.F. Filley, and R.F. Grover: Hypoxic ventilatory drive in normal man. *J.Clin.Invest.* **49**:1061-1072 (1970).

SINGLE BREATH TRACING FOR CARBON DIOXIDE IN SEPTIC PATIENTS WITH TISSUE HYPOXIA

Renzo Zatelli *

Abstract: We investigated whether tissue hypoxia in sepsis produces substantial modifications of convective airway washout and consequently of CO_2 transit time. Single breath tracing for carbon dioxide (SBT-CO_2) was analysed in 18 ICU septic patients. Nine patients had tissue hypoxia events. Using the Hill formula, all tracings were analysed point by point to obtain the time required for CO_2 to achieve 50% maximal value and the Fractional Expiratory Time 50 ($FET_{0.5}$). Hypoxic patients $FET_{0.5}$ and CO_2 clearance were compared with non-hypoxic patients data. In hypoxic group $CvCO_2$, CO_2 clearance and $FET_{0.5}$ values were higher than in non hypoxic group. During the recovery from hypoxia capnographic parameters did not differ from those recorded in the hypoxic period. CO_2 clearance, but not $FET_{0.5}$, correlated with arterial lactate and base excess either in hypoxic or in recovery period. In conclusion in septic patients tissue hypoxia influences CO_2 elimination, modifying SB-CO_2 tracing and lengthening $FET_{0.5}$.

1. INTRODUCTION

Tissue hypoxia is a condition in which the cells of a tissue have abnormal oxygen utilization such that the tissue experiences anaerobic metabolism[1]. In sepsis, tissue hypoxia may be caused by pulmonary failure due to a rapid and selective accumulation of rigid neutrophils in the lungs[2]. Furthermore, in sepsis the capillary shunting can interfere not only with O_2 kinetics, but also with CO_2 elimination kinetics. However, CO_2 clearance is measured by mixed venous CO_2 determination by means of pulmonary artery catheterisation or rebreathing methods. Since capnography in ICU is extensively employed in critically ill patients, it would be useful to obtain a relationship between the capnographic parameters and tissue hypoxia. Although the most important practical use of capnography is the early warning of catastrophic events, single breath tracing for CO_2 (SBT-CO_2) analysis has been used to analyse more sophisticated aspects of respiratory physiology. We choose to study specific aspects of SB-CO_2 tracing, due to the strong correlations between spirometry and capnography observed analysing the intermediate phase.[3]

* Renzo Zatelli, Dept. of Anesthesia and Intensive Care, University of Ferrara, 44100 Ferrara, Italy

The purpose of our study was to verify the hypothesis that in sepsis SB-CO_2 tracing is different in hypoxic (H) compared with non hypoxic (NH) patients, and that tracing modifications correlate with markers of impaired O_2 utilization.

To test our hypothesis we performed an observational study on septic patients admitted to the general ICU of the University Hospital of Ferrara-Italy.

2. PATIENTS AND METHODS

Eighteen ICU septic patients were studied[4]. All patients had a pulmonary artery catheter for determination of cardiac output, pulmonary arterial pressures, and mixed venous gasanalysis. Besides usual laboratory analysis, arterial lactate (LACT) and base excess (BE) were serially performed. O_2 consumption index (VO_2I), CO_2 elimination index (VCO_2I) and respiratory quotient (RQ) were continuously monitored with an indirect calorimetric device. Subjects were mechanically ventilated in control mode with constant inspiratory flow. The ventilator frequency was set to 15 breaths/min with an inspiration:expiration ratio = 1:3. Tidal volume was adjusted to maintain end-tidal CO_2 ($ETCO_2$) tension between 4.0-4.5 kPa. Inspiratory O_2 fraction was set to mantain hemoglobin O_2 saturation (measured by pulse oximetry) within a range of 95-100%. A mainstream CO_2 sensor was placed in the nearest site to the patient's airways to minimize the dead space. When cardiovascular and respiratory conditions were stable for at least 15 min, SB-CO_2 tracings, hemodynamic data and mixed venous blood gasanalysis were obtained. The SB-CO_2 tracings were stored by a special computer program for subsequent processing. O_2 debt was calculated according to LACT and BE[5]. All measurements were performed at admission, during hypoxia periods and at the end of the hypoxic event. Septic patients were considered as hypoxic (H) when diplaying the following characteristics[6] : LACT >2.5 mmol/L, BE < - 2.5 mEq/L and mixed venous hydrogen concentration ($v[H^+]$) >45 nanomol/L. Septic patients not meeting these criteria were considered as non hypoxic (NH).

Data sets obtained in H patients were compared with the data of NH patients. In the H group data obtained during hypoxia were compared to data recorded after recovery. The SB-CO_2 tracing is subdivided in phase I (inspiratory baseline), phase II (expiratory upstroke), phase III (expiratory or alveolar plateau), and phase IV (inspiratory downstroke)[7]. The data from central phase II enabled us to calculate the fractional expiratory CO_2 partial pressure at specific time points. A regression was performed, and intercept and slope were used in Hill equation:

$$\log ((PECO_2(t)/PACO_2)/(1 - (PECO_2(t)/PACO_2))) = a + b \log t$$

for $PECO_2(t) = 0.5 \times PACO_2 \rightarrow 0 = a + b \log t_{0.5} \rightarrow t_{0.5} = 10^{-a/b}$ (time required to reach 50% maximal CO_2 partial pressure). Dividing $t_{0.5}$ by the expiratory time we obtained the Fractional Expiratory Time ($FET_{0.5}$).

Differences between groups were compared by the Mann Whitney U test. Correlations between capnographic parameters and metabolic correlates of O_2 deficit were analyzed by linear regression. A p value <0.05 was considered as significant. Data were expressed as the mean ± SEM.

3. RESULTS

Nine patients met the H criteria and nine did not (NH). There was no significant difference between H and NH patients in age (53.5 ± 5.4 vs. 60.5 ± 4.9 years; p=0.4), weight (76.5 ± 7.3 vs. 69.4 ± 4.9 kg; p=0.4), sex distribution, and ventilation time. Illness severity scores, both APACHE II (20.0 ± 0.9 vs. 16.4 ± 0.6; p=0.0032) and SOFA[8] score (11.7 ± 0.6 vs. 9.7 ± 0.4; p=0.01) were significantly different. The hospital mortality rate in the H group was 100% compared with 44.4% in the NH group. LACT (4.5 ± 0.6 vs. 2.0 ± 0.2 mmol/L; p=0.0002), BE (-4.1 ± 1.0 vs. 0.2 ± 0.4 mEq/L; p=0.0002) and v[H^+] (51.1 ± 1.9 vs. 40.0 ± 0.5 nanomol/L; p<0.0001) are the conditions used to define tissue hypoxia and therefore were different in the two groups. In some cases among the NH group, LACT or BE values were above the cut-off limits, but the other variables were not. On the contrary, in the NH group v[H^+] never reached values >50 nanomol/L. The Pa/FIO_2 ratio (kPa/%) in the H group (33.6 ± 2.7) was significantly (p<0.0001) lower than in the NH group (50.4 ± 1.5). VO_2I (mmol/min/m^2) was not different in the two groups (5.6 ± 0.2 vs. 6.0 ± 0.2; p=0.167). Cardiac index was not significantly reduced (4.4 ± 0.2 vs. 4.9 ± 0.2 L/min/m^2; p=0.051) in H patients. The influence of hypoxia on left ventricular stroke work was quite strong (p< 0.0001), since the latter was reduced to nearly 50% in H patients (31.1 ± 2.4*0.01 J/b/m^2) in comparison with NH (51.0 ± 1.9). The indexed O_2 availability was reduced in hypoxia (23.9 ± 1.1 mmol/min/m^2 vs. 28.4 ± 0.8; p=0.0015) and O_2 extraction rate was proportionally increased (0.25 ± 0.01 vs. 0.22 ± 0.01; p=0.024). O_2 deficit, calculated on the basis of LACT and BE, was nearly zero (0.14 ± 0.08) in NH patients and 1.28 ± 0.26 mmol/min/m^2 in H patients (p=0.0001). $ETCO_2$ was not significantly reduced in hypoxia (3.9 ± 0.1 kPa vs. 4.1 ± 0.1; p= 0.115) and CO_2 elimination was similar in the two groups (6.61 ± 0.24 mmol/min/m^2 and 6.60 ± 0.23; p=0.941). $CvCO_2$ was lower in H patients (22.0 ± 0.8 mmol/L) than in NH (24.8 ± 0.5; p=0.004).

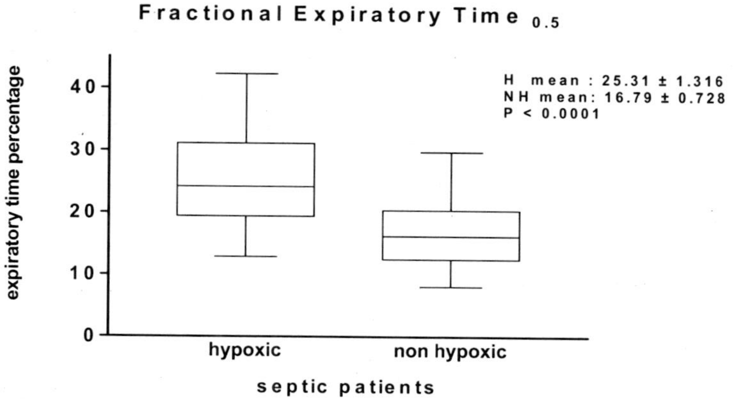

Figure 1. F.E.T.$_{0.5}$ in hypoxic (H) and non hypoxic (NH) septic patients.

CO_2 clearance was greater in H patients (0.327 ± 0.023 L/min/m^2) than in NH (0.271 ± 0.01; p=0.028). Tissue hypoxia caused a smaller α angle (72.99° ± 1.44 vs. 76.24° ± 0.74; p=0.03) and a wider β (12.09° ± 1.01 vs. 7.56° ± 0.38; p=0.0001) and γ (119.1° ± 2.02 vs. 111.3° ± 0.86; p=0.0001) angles in SB-CO_2 tracings. $t_{0.5}$ was similar in the two groups (395.0 ± 25.0 and 396.2 ± 18.1), but due to different expiratory times (1639 ± 105 and 2470 ± 102 millisecs; p< 0.0001), $FET_{0.5}$ was bigger in the H group (25.3 ± 1.3) than in the NH (16.8 ± 0.7;p<0.0001)(Fig 1).

The comparison between data obtained during the hypoxic period and after its recovery in septic hypoxic patients is shown in Table 1.

After recovery from hypoxia, LACT, BE, and v[H^+] values returned to the normal range, as well as O_2 deficit. Pa/FIO_2 remained low, while VO_2I was not different in the recovery from the hypoxic period, as well as O_2 delivery and O_2 extraction. No difference was detectable in cardiac index, while LVSWI reached the values of a non-hypoxic situation. Among respiratory parameters, $CvCO_2$ was significantly higher in the recovery period (26.48 ± 0.68 mmol/L vs. 19.46 ± 0.90, p <0.0001). VCO_2I, $ETCO_2$ and

RQ showed mild increases. No difference was detected in CO_2 clearance. Parameters derived from CO_2 tracing (angles, $t_{0.5}$ and $FET_{0,5}$) were not significantly different before and after recovery from hypoxia. In septic NH patients tissue hypoxia markers correlated with β and γ angles (Fig. 2) and $CvCO_2$.

Table 1. Parameters obtained during the hypoxic period and after its recovery.

Variable	during hypoxia	after recovery	p value
Samples number	26	15	
LACT (mmol/L)	6.31 ± 0.69	1.29 ± 0.13	< 0.0001
BE (mEq/L)	- 7.92 ± 0.89	2.55 ± 0.79	< 0.0001
V $[H^+]$ (nanomol/L)	57.59 ± 1.81	39.82 ± 1.662	< 0.0001
Pa / F I O_2 (kPa / %)	31.91 ± 3.66	36.47 ± 3.61	0.417
VO2 index (mmol/min/m^2)	5.78 ± 0.31	5.33 ± 0.28	0.335
Cardiac index (L/min/m^2)	4.38 ± 0.30	4.35 ± 0.30	0.951
L.V.S.W.I. (0.01 J/b/m^2)	24.30 ± 1.89	42.93 ± 4.24	< 0.0001
O_2 delivery (mmol/min/m^2)	24.32 ± 1.33	23.24 ± 1.95	0.648
O_2 extraction	0.250 ± 0.017	0.248 ± 0.020	0.945
O_2 deficit (mmol/min/m^2)	2.20 ± 0.28	0.30 ± 0.10	< 0.0001
VCO_2 index (mmol/min/m^2)	11.64 ± 0.7	13.85 ± 0.42	0.0136
respiratory quotient	1.14 ± 0.08	1.42 ± 0.09	0.0269
ET CO_2 (kPa)	3.71 ± 0.15	4.20 ± 0.19	0.0498
Mean $CvCO_2$ (mmol/L)	19.46 ± 0.90	26.48 ± 0.68	< 0.0001
CO_2 clearance (L/min/m^2)	0.355 ± 0.034	0.278 ± 0.015	0.106
α angle (degrees)	72.16 ± 1.89	74.44 ± 2.21	0.452
β angle (degrees)	12.49 ± 1.31	11.39 ± 1.63	0.606
γ angle (degrees)	120.3 ± 2.64	117.0 ± 3.15	0.431
Time $_{0.5}$ $PACO_2$ (milliseconds)	389.8 ± 30.6	404.1 ± 44.3	0.786
expiratory time (milliseconds)	1614 ± 112.2	1682 ± 218.1	0.762
F.E.T. $_{0.5}$ (%)	24.94 ± 1.74	25.97 ± 2.04	0.710

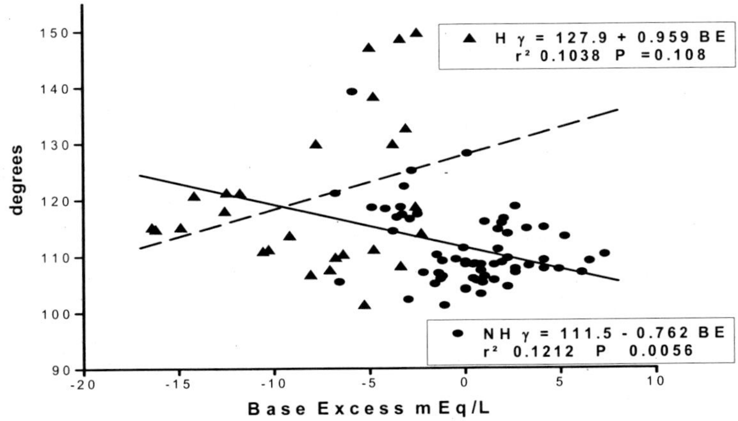

Figure 2. SB-CO2 γ angle in hypoxic and non hypoxic patients.

In septic patients with hypoxic periods metabolic correlates of O_2 debt correlated with CO_2 clearance, $CvCO_2$, $ETCO_2$, VCO_2I, LVSWI, and $t_{0.5}$. In the acute phase of hypoxia, BE and LACT correlated with the latter parameters plus RQ, but not with $t_{0.5}$. In the recovery period, hypoxia markers correlate only with $CvCO_2$ and CO_2 clearance.

4. DISCUSSION AND CONCLUSION

The presence of ongoing O_2 debt is deleterious, resulting in an inflammatory cascade that can lead to multisystem organ disfunction. The rapid identification and restoration of the O_2 debt are central to the resuscitation of the critically ill patient[9]. There is no perfect indicator of dysoxia[1,10], but it is known that perfusion failure represents a dual defect of tissue O_2 deficit and CO_2 excess[11], and venous hypercarbia is associated with systemic hypoperfusion and severe sepsis. In comparison with CO_2 kinetics parameters, other O_2 deficit markers are more invasive, time-consuming and expensive. $ETCO_2$ represents an indirect measurement of pulmonary blood flow only when cardiac output is profoundly reduced[12]. Recently, tissue PCO_2, evaluated by sublingual measurement, was considered as highly predictive of outcome[13].

SB-CO_2 tracing was defined as the respiratory equivalent of electrocardiography, and could be an easy and useful means to detect tissue hypoxia by modifications of its shape[14]. Many attempts have been made to standardize the capnographic tracing[15] and computer algorithms were built up for classifying capnograms[16]. Our results further demonstrate the usefulness of such analysis. Our patients were ventilated with different PEEP, which was set to zero when SB-CO_2 tracings were performed, because in this condition a relationship between disease severity and capnographic parameters was detected[17]. In our observations tissue hypoxia, as well as ARDS[18], induced an increase in Phase III slope (from 0.13 to 0.22) and VD/VT (from 0.39 to 0.46) and a reduction in the Relative Peripheral Shunt Index (from 0.91 to 0.24). In our septic patients the $CvCO_2$ decreased due to hypoxia and increased in the recovery period. This is in agreement with the proton formation during anaerobic metabolism and subsequent reaction with HCO_3^- to form CO_2 and water. Since the blood supply is severely limited during a hypoxic period, CO_2 is not transported away and accumulates in the tissue. Consequently, the more ischemic the tissue, the less reliable is a venous blood sample in reflecting the condition of the tissue[19].

In conclusion in septic patients tissue hypoxia influences CO_2 elimination from alveoli to the atmosphere, modifying SB-CO_2 tracing and lengthening $FET_{0.5}$.

Capnography indices detect important distribution abnormalities in septic hypoxic patients compared to septic non hypoxic subjects during mechanical ventilation. The capnographic indices best related to O_2 deficit markers were β and γ angles, but during the acute phase of hypoxia this correlation was no longer present. Capnographic changes due to tissue hypoxia tend to persist, while circulatory and oxygen transport changes rapidly recover, as the increase in lung and airway resistances due to hypoxia return within the normal range more slowly than cardiovascular dynamics.

5. REFERENCES

1. F.Lemaire, G.Apolone, L.Blanch, J.Y.Fagon, P.Foex, G.Hedenstierna, D. Morel, P.Parsons, A.Pesenti, G.Ramsay, G.Simbruner, and A.S.Slutsky, Tissue hypoxia. How to detect, how to correct, how to prevent?, *Intensive Care Med* **22**(5),1250-1257 (1996).
2. B.Heit, and P.Kubes, Bacterial products sabotage the innate immune system, *Advances in sepsis* **3**(4), 114-120 (2004).
3. L.Blanch, R.Fernandez, P.Saura, F.Baigorri, and A.Artigas, Relątionship between expired capnogram and respiratory system resistance in critically ill patients during total ventilatory support, *Chest* **105**(1), 219-223 (1994).

4. R.C.Bone, R.A.Balk, and F.B.Cerra, Definitions for sepsis and organ failure and guidelines for the use of innovative therapies in sepsis, *Chest* **101**(7),1644-1655 (1992).
5. C.M.Dunham, J.H.Siegel, L.Weireter, M.Fabian, S.Goodarzi, P.Guadalupi, L.Gettings, S.E.Linberg, and T.Vary, Oxygen debt and metabolic acidemia as quantitative predictors of mortality and the severity of the ischemic insult in hemorrhagic shock, *Crit Care Med.* **19**(2),231-243 (1991).
6. D. Rixen and J.H. Siegel, Metabolic correlates of oxygen debt predict posttrauma early acute respiratory distress syndrome and the related cytokine response, J Trauma **49** (3), 392 – 403 (2000).
7. S.Hussain, and D.T.Rafael, Analysis of a straight-line capnographic waveform, *Anest Analg* **85** (2), 465-474 (1997).
8. J.L.Vincent, R.Moreno, and J.Takala, The SOFA (Sepsis-related Organ Failure Assessment) score to describe organ dysfunction/failure, *Intensive Care Med.* **22**(3),707-710 (1996).
9. R.N.Bilkovski, E.P.Rivers, and H.M. Horst, Targeted resuscitation strategies after injury, *Curr Opin Crit Care* **10**(6), 529-538 (2004)
10. J.A.Kruse, Searching for the perfect indicator of dysoxia. *Crit Care Med* **27**(3), 469-471 (1999).
11. M.H.Weil, Tissue PCO_2 as universal marker of tissue hypoxia. *Minerva Anestesiol* **66**(5), 343-347 (2000).
12. X.Jin, M.H.Weil, W.Tang, H.Povoas, A.Pernat, J.Xie, and J.Bisera, End-tidal carbon dioxide as a noninvasive indicator of cardiac index during circulatory shock, *Crit Care Med* **28**(7), 2415-2419 (2000).
13. P.E.Marik, and A.Bankov, Sublingual capnometry versus traditional markers of tissue oxygenation in critically ill patients, *Crit Care Med.* **31**(3), 818-822 (2003).
14. R.Fletcher, *The single breath test for carbon dioxide* (Berlings Arlov, Lund, Sweden 1986).
15. J.L.Benumof, Interpretation of capnography, *AANA J* **66**(2),169-176(1998).
16. H.R.van Genderingen, N.Gravenstein, J.J. van der Aa, and J.Gravenstein, Computer-assisted capnogram analysis, *J Clin Monit* **3**(3), 194-200 (1987).
17. L.Blanch, U.Lucangelo, J.Lopez-Aguilar, R.Fernandez, and P.V.Romero, Volumetric capnography in patients with acute lung injury: effects of positive end-expiratory pressure. *Eur Respir J.* **13**(5), 1048-1054 (1999).
18. P.V.Romero, U.Lucangelo, J.Lopez-Aguilar, R.Fernandez, and L.Blanch, Physiologically based indices of volumetric capnography in patients receiving mechanical ventilation. *Eur Respir J.* **10**(6), 1309-1315 (1997).
19. T.I.Tonnessen, Biological basis for PCO_2 as a detector of ischemia, *Acta Anaesthesiol Scand* **41**(5), 659-669 (1997).

AUTHOR INDEX

SUBJECT INDEX

Printed in the USA